Chemical/Instrumental Approaches to the Evaluation of Wine Chemistry

Chemical/Instrumental Approaches to the Evaluation of Wine Chemistry

Special Issue Editors

Rosa Perestrelo

José S. Câmara

MDPI • Basel • Beijing • Wuhan • Barcelona • Belgrade • Manchester • Tokyo • Cluj • Tianjin

Special Issue Editors
Rosa Perestrelo
Universidade da Madeira
Portugal

José S. Câmara
Universidade da Madeira
Portugal

Editorial Office
MDPI
St. Alban-Anlage 66
4052 Basel, Switzerland

This is a reprint of articles from the Special Issue published online in the open access journal *Molecules* (ISSN 1420-3049) (available at: https://www.mdpi.com/journal/molecules/special_issues/instrumental_wine_chemistry).

For citation purposes, cite each article independently as indicated on the article page online and as indicated below:

LastName, A.A.; LastName, B.B.; LastName, C.C. Article Title. *Journal Name* **Year**, *Article Number*, Page Range.

ISBN 978-3-03928-734-5 (Hbk)
ISBN 978-3-03928-735-2 (PDF)

Contents

About the Special Issue Editors

Rosa Perestrelo (Ph.D.) is a Researcher at Madeira University and is member of the Madeira Chemistry Research Center (CQM) research unit. She studied chemistry at Madeira University, completed a M.S. degree in green chemistry (Faculty of Sciences of Lisbon University), and finished her Ph.D. in 2013, where she evaluated the potential of Vitis vinifera L. grapes used to produce Madeira wine. She has published 53 SCI papers, 6 book chapters, and more than 90 poster presentations in international and national conferences and about 20oOral communications. H-index: 17; Number of citations: 1157, Scopus; ORCID No.: 0000-0002-7223-1022.

José S. Câmara (Ph.D.) received his M.S. degree in quality control in chemistry in 1997 at the University of Coimbra. In 2004, he obtained his Ph.D. in chemistry (analytical chemistry) at University of Madeira. Between 1993 and 2004, he worked at University of Madeira as a lecturer and as researcher with Professor José Carlos Marques on the Analytical Characterization of Madeira wines. In 2004, he joined the University of Madeira as Assistant Professor where he started his independent research on food chemistry and, afterward, on medicinal chemistry. Since 2004, he has been a senior member at CQM where he is currently developing his research. Since 2016, he has been a member of the executive committee of CQM and he is responsible of the Natural Products Research group of CQM. He is a supervisor of Ph.D. and M.S. thesis and Director of the Course of Chemistry and CET on Hygiene and Food Security. He is coordinator of ERA-Net projects. José S. Câmara is coauthor of more than 100 papers (H-index = 38: i10 = 65. Google Scholar) including several reviews, encyclopedia, and book chapters, in addition to more than 350 communications in international and national conferences. He is Associate Editor of JIOMICS and member of the Editorial Board of some other journals. ORCID ID: 0000-0003-1965-3151 Researcher ID: G-3003-2013 Scopus Author ID: 10140393000 Publons: publons.com/a/1206082/.

Preface to "Chemical/Instrumental Approaches to the Evaluation of Wine Chemistry"

Wine is a widely consumed beverage due to its unique and pleasant sensory properties. Wine is composed of more than 1000 chemical compounds (e.g., alcohols, esters, acids, terpenoids, phenolic compounds, flavonoids, anthocyanins, minerals, and vitamins, among others) resulting from several chemical and biochemical processes. Microextraction techniques in tandem with high-resolution analytical instruments have been applied by wine researchers to expand the knowledge of wine's chemical composition with the purposes of improving wine quality, supporting winemaker decisions related to the winemaking process, and guaranteeing the authenticity of wine. As a result, we proposed "Chemical/Instrumental Approaches to the Evaluation of Wine Chemistry" as a topic for a Special Issue in Molecules. This Special Issue aims to provide an update on the state-of-the-art extraction procedures (e.g., solid-phase microextraction (SPME)) and analytical tools (e.g., nuclear magnetic resonance (NMR), inductively coupled plasma mass spectrometry (ICP-MS), and ultra-performance liquid chromatography tandem mass spectrometry (UPLC-MS/MS)), emphasizing their use as suitable platforms for the establishment of the chemical composition of wine (volatomic profile, antioxidants, phenolic pattern, elemental composition, among others). Information related to wine sensorial properties, contaminants, authenticity, and chemometric tools used for data treatment are described in this Issue. This Special Issue is accessible through the following link: https://www.mdpi.com/journal/molecules/special_issues/instrumental_wine_chemistry. As Guest Editors for this Special Issue, we would like to thank all the authors and co-authors for their contributions, all reviewers for their effort in revising the manuscripts, as well as the editorial office of Molecules for their generous help in organizing this Special Issue.

Rosa Perestrelo, José S. Câmara
Special Issue Editors

molecules

MDPI

Editorial

Chemical/Instrumental Approaches to the Evaluation of Wine Chemistry

Rosa Perestrelo * and José S. Câmara *

CQM—Centro de Química da Madeira, University of Madeira, 9020-105 Funchal, Portugal
* Correspondence: rmp@staff.uma.pt (R.P.); jsc@uma.pt (J.S.C.)

Received: 5 March 2020; Accepted: 6 March 2020; Published: 17 March 2020

Wine is a widely consumed beverage thanks to its unique and pleasant sensory properties. Wine is composed of more than one thousand chemical compounds (e.g., alcohols, esters, acids, terpenoids, phenolic compounds, flavonoids, anthocyanins, minerals, vitamins, among others) resulting from several chemical and biochemical processes [1,2]. Nowadays, microextraction techniques tandem with high-resolution analytical instruments have been applied by wine researchers to expand the knowledge of wine´s chemical composition with the purpose to improve wine quality, support winemaker decisions related to the winemaking process, and guarantee the authenticity and genuineness of wine [3–6].

As a result, we proposed "Chemical/Instrumental Approaches to the Evaluation of Wine Chemistry" as an interesting topic for a Special Issue in the Molecules journal. This Special Issue aims to update the top-of-the-art extraction procedures (e.g., solid-phase microextraction (SPME)) and analytical tools (e.g., nuclear magnetic resonance (NMR), inductively coupled plasma mass spectrometry (ICP-MS), ultra-performance liquid chromatography tandem mass spectrometry (UPLC-MS/MS)), emphasizing their use as suitable platforms for the establishment of the chemical composition of wine (volatomic profile, antioxidants, phenolic pattern, elemental composition, among others). In addition, information related to wine sensorial properties, contaminants, authenticity, and chemometric tools used for data treatment will be described in this issue. Thus, this Special Issue includes eight publications using different analytical approaches for the evaluation of wine chemistry [7–14]. Regarding gas chromatography, Sancho-Galán et al. [11] used gas chromatograph equipped with a flame ionization detector (GC-FID) to study the use of bee pollen as a flor velum activator in biological aging wines. Moreover, headspace solid-phase microextraction combined with gas chromatography-mass spectrometry (HS–SPME/GC–qMS) was used by Dang et al. [14] for determining the retention of volatile phenols (putative markers for Brettanomyces and smoke taint off-odors) by cyclodextrin in model wine, as well as by Perestrelo el al. [7] to investigate the volatile organic compounds (VOCs) that may potentially be responsible for specific descriptors of Madeira wine, providing details about Madeira wine aroma notes at the molecular level.

Related to liquid chromatography, Tarapatskyy et al. [13] used ultra-performance reverse-phase liquid chromatography tandem mass spectrometry (UPLC-MS/MS) to assess the bioactive compounds in white and red wines enriched with a *Primula veris* L. In addition, a novel and accurate method based on ultrahigh performance liquid chromatography (UHPLC) with a photo-diode array detector (PDA) and charged aerosol detector (CAD) was developed for simultaneously determining nine sweeteners (most authorized for use in China) in white spirits by Ma et al. [8].

Deng et al. [10] used inductively coupled plasma mass spectrometry (ICP-MS) to determine the concentration of trace elements in wines and health risk assessment via wine consumption was investigated in 315 wines. In this context, Tamasi et al. [9] used ionic exchange resins and hydrogels for capturing metal ions (Na, K, Mg, Ca, Mn, Fe, Cu and Zn)) in sweet dessert wines. Moreover, Li et al. [12] used a nanoparticle tracking analysis (NTA) and UV-visible spectroscopy and dynamic light scattering (DLS) to characterize the interactions between grape seed tannin and either a mannoprotein or an arabinogalactan in model wine solutions of different ethanol concentrations.

This Special Issue is accessible through the following link: https://www.mdpi.com/journal/molecules/special_issues/instrumental_wine_chemistry

As Guest Editors for this Special Issue, we would like to thank all the authors and co-authors for their contributions, all reviewers for their effort in revising the manuscripts, as well as the editorial office of Molecules journal for their generous help in organizing this Special Issue.

Funding: This research received no external funding.

Conflicts of Interest: The authors declare no conflict of interest.

References

1. Tuberoso, C.I.G.; Serreli, G.; Congiu, F.; Montoro, P.; Fenu, M.A. Characterization, phenolic profile, nitrogen compounds and antioxidant activity of Carignano wines. *J. Food Compos. Anal.* **2017**, *58*, 60–68. [CrossRef]
2. Zhao, P.; Gao, J.; Qian, M.; Li, H. Characterization of the key aroma compounds in Chinese Syrah wine by gas chromatography-olfactometry-mass spectrometry and aroma reconstitution studies. *Molecules* **2017**, *22*, 1045. [CrossRef] [PubMed]
3. Pereira, L.; Gomes, S.; Castro, C.; Eiras-Dias, J.E.; Brazão, J.; Graça, A.; Fernandes, J.R.; Martins-Lopes, P. High Resolution Melting (HRM) applied to wine authenticity. *Food Chem.* **2017**, *216*, 80–86. [CrossRef] [PubMed]
4. Epova, E.N.; Bérail, S.; Séby, F.; Vacchina, V.; Bareille, G.; Médina, B.; Sarthou, L.; Donard, O.F.X. Strontium elemental and isotopic signatures of Bordeaux wines for authenticity and geographical origin assessment. *Food Chem.* **2019**, *294*, 35–45. [CrossRef] [PubMed]
5. Gougeon, L.; da Costa, G.; Richard, T.; Guyon, F. Wine Authenticity by Quantitative 1H NMR Versus Multitechnique Analysis: A Case Study. *Food Anal. Methods* **2019**, *12*, 956–965. [CrossRef]
6. Kokkinofta, R.; Fotakis, C.; Zervou, M.; Zoumpoulakis, P.; Savvidou, C.; Poulli, K.; Louka, C.; Economidou, N.; Tzioni, E.; Damianou, K.; et al. Isotopic and elemental authenticity markers: A case study on Cypriot wines. *Food Anal. Methods* **2017**, *10*, 3902–3913. [CrossRef]
7. Perestrelo, R.; Silva, C.; Câmara, J.S. Madeira wine volatile profile. A platform to establish madeira wine aroma descriptors. *Molecules* **2019**, *24*, 3028. [CrossRef] [PubMed]
8. Ma, K.; Li, X.; Zhang, Y.; Liu, F. Determining high-intensity sweeteners in white spirits using an ultrahigh performance liquid chromatograph with a photo-diode array detector and charged aerosol detector. *Molecules* **2020**, *25*, 40. [CrossRef] [PubMed]
9. Tamasi, G.; Pardini, A.; Bonechi, C.; Donati, A.; Casolaro, M.; Leone, G.; Consumi, M.; Cini, R.; Magnani, A.; Rossi, C. Ionic exchange resins and hydrogels for capturing metal ions in selected sweet dessert wines. *Molecules* **2018**, *23*, 2973. [CrossRef] [PubMed]
10. Deng, Z.H.; Zhang, A.; Yang, Z.W.; Zhong, Y.L.; Mu, J.; Wang, F.; Liu, Y.X.; Zhang, J.J.; Fang, Y.L. A human health risk assessment of trace elements present in Chinese wine. *Molecules* **2019**, *24*, 248. [CrossRef] [PubMed]
11. Sancho-Galán, P.; Amores-Arrocha, A.; Jiménez-Cantizano, A.; Palacios, V. Use of multiflora bee pollen as a flor velum yeast growth activator in biological aging wines. *Molecules* **2019**, *24*, 1763. [CrossRef] [PubMed]
12. Li, S.; Wilkinson, K.L.; Mierczynska-Vasilev, A.; Bindon, K.A. Applying nanoparticle tracking analysis to characterize the polydispersity of aggregates resulting from tannin–polysaccharide interactions in wine-like media. *Molecules* **2019**, *24*, 2100. [CrossRef] [PubMed]
13. Tarapatskyy, M.; Kapusta, I.; Gumienna, A.; Puchalski, C. Assessment of the bioactive compounds in white and red wines enriched with a primula veris L. *Molecules* **2019**, *24*, 4074. [CrossRef]
14. Dang, C.; Wilkinson, K.L.; Jiranek, V.; Taylor, D.K. Development and evaluation of a HS-SPME GC-MS method for determining the retention of volatile phenols by cyclodextrin in model wine. *Molecules* **2019**, *24*, 3432. [CrossRef]

molecules

MDPI

Article

Ionic Exchange Resins and Hydrogels for Capturing Metal Ions in Selected Sweet Dessert Wines

Gabriella Tamasi [1,2,*], Alessio Pardini [1,2], Claudia Bonechi [1,2,*], Alessandro Donati [1,2], Mario Casolaro [1], Gemma Leone [1,3], Marco Consumi [1,3], Renzo Cini [1], Agnese Magnani [1,3] and Claudio Rossi [1,2,4]

1 Department of Biotechnology, Chemistry and Pharmacy, University of Siena, Via Aldo Moro 2, 53100 Siena, Italy; pardini4@student.unisi.it (A.P.); alessandro.donati@unisi.it (A.D.); mario.casolaro@unisi.it (M.C.); gemma.leone@unisi.it (G.L.); marco.consumi@unisi.it (M.C.); renzo.cini@unisi.it (R.C.); agnese.magnani@unisi.it (A.M.); claudio.rossi@unisi.it (C.R.)
2 Centre for Colloid and Surface Science (CSGI), University of Florence, Via della Lastruccia 3, 50019 Sesto Fiorentino, Firenze, Italy
3 National Interuniversity Consortium of Materials Science and Technology (INSTM), Via G. Giusti 9, 50121 Firenze, Italy
4 Operative Unit, University of Siena, CampoVerde, Calabria, Italy
* Correspondence: gabriella.tamasi@unisi.it (G.T.); claudia.bonechi@unisi.it (C.B.); Tel.: +39-0577-233780 (G.T.); +39-0577-232123 (C.B.)

Academic Editors: Rosa Perestrelo and José Sousa Câmara
Received: 18 October 2018; Accepted: 11 November 2018; Published: 14 November 2018

Abstract: Samples of sweet and dessert wines, Vin Santo (VSR) from Malvasia grapes, and Granello (GR) from Sauvignon grapes were collected and analyzed for the content of selected macro- and micro-nutrients (Na, K, Mg, Ca, Mn, Fe, Cu and Zn) and of Pb. GR wines had low levels for Fe, Cu and Zn, when compared to VSR and in particular Zn was two orders of magnitude lower. Methods to decrease the content of Zn and Cu in VSR, as well as those for reducing, at the same time, the concentrations of Ca, Mg and K in both VSR and GR, to avoid the formation of opalescence and depots of metal tartrates, were studied. Synthetic hydrogels containing L-histidine residue were tested. The overall relative lowering effects were by *ca* 4, 23, and 12% for K, Mg and Ca contents, and *ca* 6, 27 and 10%, for Mn, Cu and Zn contents, in GR wine samples. Commercial ion exchange resin Lanxess Lewatit L-207 and L-208 were then assayed, being legally allowed in the agro-food industry. The L-207 resin revealed great lowering effects on the concentrations of Mn, Cu and Zn, being 75, 91 and 97%, respectively, in VSR wines and 77, 76 and 92%, respectively, in GR wines. The content of Zn was reduced from 49.3 ± 1.2 mg/L in the original wine, down to 1.1 ± 0.1 mg/L, within 48 h soaking. The effects on the character of the dessert wines by the resin L-207 was also taken under control, measuring pH and color index. The color index changed by *ca* 15% and pH by *ca* 6% upon treatment of VSR wine with L-207 resins (48 h).

Keywords: dessert wine; ionic exchange resin; hydrogels; metals; atomic absorption

1. Introduction

Vin Santo (VSR) is a well known dessert wine produced and used mostly in Tuscany region (Italy) for several centuries, but even other regions in Italy contribute to the production of this beverage [1]. Furthermore, the European Community regulation by the year 2005 admits the usage of the term for certain dessert wine from the Santorini island in Greece [2]. It is obtained from grapes that, after harvesting, are aged for some two months in dark, well aerated and dustless rooms, at *ca* 18–20 °C. That is usually performed by storing the grapes over racks of layers of reeds, or hanging the bunches with wood or metal racks, so that the berries can lose most of the humidity content (drying step), without developing any significant amount of mold. The control of the main chemical parameters (like, pH, overall acidity) and of the stability of the product during the wine-making stages (refinement, storing and aging processes), are of great importance for preserving the quality of final products, as well as for economical aspects of the wine-making companies, and even for the territory. An amber color (for Vin Santo) or pale yellow (for Granello (GR), "passito" wine), excellent clearness, good brilliance, and a fresh and characteristic fruity smell are related to low pH values. Even the concentrations of cationic micro- and macro-nutrients like iron, copper and zinc, as well as potassium and calcium, and specific ranges for concentrations of anionic species, play important roles for the quality of those wines. It is well known that pH higher than 3.8–4.0 negatively influence on the color and aroma, and decreases the chemical and microbiological stability of wines [3–6]. Moreover, it is known, that wines at high pH cannot be corrected by addition of acids (i.e., tartaric acid, succinic acid) for several reasons: for example, the wine could already have a high tartaric acidity, and further addition of organic acid could allow the formation of potassium hydrogen tartrate (KHT) deposits, especially if the content of potassium is also high. Another undesired effect caused by high concentrations of tartaric acid is the sense of astringency. The usage of ionic exchange resins for the pH and acidity tuning, as well as for improving the tartaric and oxidative stability in wines, has been performed for decades, by using many kind of materials, spanning from commercial styrene-divinylbenzene copolymer gels, mostly containing functional groups of sulphonic acid [7–10], polyvinylimidazole and polyvinylpyrrolidone copolymers (PVI-PVP) [11,12], Dowex resins [13], chelating iminodiacetate [13], and carboxylmethylcellulose (CMC) [14] based materials. However, the resins that are commonly used by the wine-making companies, and currently accepted by law [15], are only PVI-PVP copolymers and CMC-based materials. In the case of the common divinylbenzene family resins, the supporting skeleton is functionalized with groups having high affinity towards cations and anions (i.e., if the resin is of the general type R ... H^+ and has good affinity for K^+, Mg^{2+} and Ca^{2+} ions, it can be used to keep under control the content of those cations). Therefore, once the wine is passed through a column packed with, or stored on a bed of that type of resin, the liquid will experience a lowering of K^+, Mg^{2+} and Ca^{2+} concentrations (and then a decreased risk for precipitation of corresponding metal hydrogen tartrate), but will also experience a significant lowering of pH that cannot be below 2.8. Furthermore, those processes might cause damage to certain wine components that contribute to defining the aroma, the color, or that are very important for the nutraceutical potentials (resveratrol, quercetin). These parameters have been considered and reported [8,9,13]. The ion exchange resins also need regeneration over time, for instance the general resin R ... H^+ could be regenerated by a solution of sulfuric acid or hydrochloric acid, at a proper concentration, and the liquid that comes from the regeneration processes should be properly treated before it can be disposed of. Moreover, the degradation and erosion of the resin during the treatment of the wine must be carefully monitored, in order to avoid significant pollution of the wine, by the species from the resin itself. Those methodologies are approved and currently performed in wine-making and refining procedures [4,5]. This work aimed also to investigate for the first time the usage of synthetic materials based on temperature-responsive hydrogels, carrying amphoteric units of L-histidine residues able to form complex heavy metal ions. This part of the work came out from the experience acquired by the research group on drugs, metal-based-drugs, nutraceuticals properties [16–22] and delivery systems, for embedding and releasing bioactive molecules, such as smart hydrogels, cross-linked polysaccharides, and liposomes [23–26].

2. Results

2.1. The Content of Selected Nutrient Metals in Granello (GR) and Vin Santo (VSR) Dessert Wines

The trend of the content for macronutrients C_{Na}, C_K, C_{Mg}, and C_{Ca} in GR wine, during the time period 58th–110th days, from the sample collections (from the oak barrels at wine making company;

Table 1, Figure S1a,b) showed that after 58 days in the laboratory conditions, the GR wine could be considered stabilized. The contents of the alkali metals C_{Na}, C_K, C_{Mg} and C_{Ca} averaged 28.3 ± 0.3, 754 ± 17, 200 ± 6, and 60 ± 3 mg/L at plateau underwent a significant decrease with respect to the initial values (7th day for GR samples; Table 1). The formation of opalescence and precipitate occurred during the stabilization time (before the 58th day), because the formation of insoluble species containing Na, K, and Ca. In fact, the concentrations of those metals decreased by 49, 11, and 66%, respectively. The contents of metal micronutrients C_{Mn}, C_{Fe}, C_{Cu} and C_{Zn} (Table 1, Figure S1c) revealed a small decrease for Mn (5%) down to the plateau value by 1.56 ± 0.02 mg/L, a decrease by at least 29% for Fe (0.28 ± 0.03 mg/L to below the limits of quantification (LOQ), 0.20 mg/L) and slight increases for Cu (15%) and Zn (2%) at plateau values by 0.39 ± 0.01 and 0.58 ± 0.02 mg/L, respectively. By contrast, the C_{Mg} increased significantly (*ca* 117%) from starting value (92 ± 7 mg/L) up to plateau value by 200 ± 6 mg/L. It can be noticed that the original sample bottles showed deposits and haziness, and during the two months in laboratory conditions, the entity of these phases changed markedly. That behavior would be reasonably interpreted as the occurring of chemical and biochemical processes towards an equilibrium state, meaning that some elements can increase and other decrease the total content in the final stabilized wine, on the basis of the specific solubility/stability, initial amount, kinetic of enzymatic reaction, of the single species. It is reasonable that Mg, which is an important co-factor for many types of enzymatic reactions, undergoes relevant changes.

Table 1. Content of metal nutrients (C_{Na}, C_K, C_{Mg}, C_{Ca}, C_{Mn}, C_{Cu}, C_{Zn}, as mg/L) in GR wines as function of time (days). Values for C_{Fe} < LOQ (0.20 mg/L).

Day	C_{Na}	C_K	C_{Mg}	C_{Ca}	C_{Mn}	C_{Cu}	C_{Zn}
58th	28.4 ± 0.5	770 ± 22	205 ± 3	64 ± 5	1.59 ± 0.05	0.40 ± 0.02	0.58 ± 0.03
59th	28.6 ± 0.5	768 ± 19	200 ± 3	65 ± 4	1.58 ± 0.04	0.39 ± 0.03	0.58 ± 0.03
60th	28.2 ± 0.5	767 ± 19	206 ± 6	66 ± 3	1.60 ± 0.03	0.38 ± 0.02	0.58 ± 0.03
61th	28.1 ± 0.5	765 ± 18	201 ± 5	63 ± 5	1.56 ± 0.02	0.40 ± 0.03	0.59 ± 0.04
62th	28.0 ± 0.6	730 ± 16	201 ± 3	60 ± 3	1.55 ± 0.02	0.39 ± 0.03	0.57 ± 0.04
63th	28.1 ± 0.6	738 ± 15	207 ± 5	63 ± 4	1.57 ± 0.02	0.41 ± 0.01	0.59 ± 0.05
64th	28.0 ± 0.5	753 ± 10	204 ± 6	60 ± 3	1.56 ± 0.02	0.40 ± 0.04	0.58 ± 0.03
65th	28.2 ± 0.6	760 ± 23	194 ± 5	59 ± 3	1.54 ± 0.02	0.37 ± 0.03	0.60 ± 0.03
66th	28.3 ± 0.3	775 ± 11	192 ± 7	62 ± 2	1.56 ± 0.03	0.39 ± 0.01	0.59 ± 0.03
67th	28.0 ± 0.6	740 ± 14	186 ± 5	58 ± 3	1.56 ± 0.03	0.40 ± 0.01	0.63 ± 0.05
68th	28.3 ± 0.6	725 ± 14	202 ± 4	55 ± 2	1.56 ± 0.02	0.40 ± 0.02	0.58 ± 0.04
70th	28.4 ± 0.5	735 ± 14	197 ± 5	59 ± 2	1.57 ± 0.02	0.37 ± 0.03	0.56 ± 0.04
80th	28.7 ± 0.7	760 ± 14	199 ± 5	58 ± 2	1.55 ± 0.03	0.39 ± 0.03	0.60 ± 0.02
90th	28.4 ± 0.6	750 ± 15	202 ± 4	57 ± 5	1.54 ± 0.02	0.40 ± 0.03	0.59 ± 0.03
100th	29.3 ± 0.6	755 ± 12	200 ± 4	55 ± 3	1.56 ± 0.02	0.36 ± 0.03	0.57 ± 0.03
110th	28.5 ± 0.6	780 ± 17	210 ± 8	57 ± 6	1.56 ± 0.02	0.37 ± 0.02	0.57 ± 0.04
Average *	28.3 ± 0.3	754 ± 17	200 ± 6	60 ± 3	1.56 ± 0.02	0.39 ± 0.01	0.58 ± 0.02
7th	55.7 ± 3.6	847 ± 50	92 ± 7	177 ± 7	1.65 ± 0.08	0.34 ± 0.04	0.57 ± 0.05
Δ% *	−49	−11	117	−66	−5	15	2

Values (mean ± ESD) obtained from three samples, and three replicates each sample; full experimental data sets are reported in Table S1. * Average values calculated from 58th–110th data; Δ% Relative percentage change = [(Average (58th − 110th) − 7th)/7th] × 100.

The macronutrients data relevant to Vin Santo dessert wines (VSR) revealed a general trend similar to that, just above commented for GR wine (Table 2, Figure S2 and Table S2). In fact, the C_{Na}, C_K, and C_{Ca} in VSR wines underwent a general decrease during the stabilization time (before the 58th day), down to plateau averaged values, by 20.0 ± 0.5, 685 ± 24, 41 ± 4 mg/L, thus the lowering effects were 43, 12, and 32%, respectively, when based on initial values (7th day, 35.0 ± 0.5, 780 ± 28 and 61 ± 5 mg/L for C_{Na}, C_K and C_{Ca}, respectively). The most interesting data for micronutrients were those relevant to the content of Zn and Cu. In fact, their magnitudes at plateau 49.3 ± 1.2 (C_{Zn}) and 2.68 ± 0.03 (C_{Cu}) mg/L were much higher than for GR wines, and were also *ca* 25–100 times the values found for high-quality red wines from the same company [27], and higher than the maximum allowable concentrations by European and Italian laws (MAC_{Zn}, 5 mg/L, MAC_{Cu}, 1 mg/L) [28,29]). Even the C_{Mn}, and C_{Fe} at plateau, were higher than those found for GR. Those facts had a rationale in the habit to use metal devices (iron wires or zinc-copper alloy wires) to hang up the bunches at the drying stage for the preparation of VSR. Beside those observations, the absolute values for C_{Mn} and C_{Fe} were below the maximum allowable concentrations by import regulation from China (MAC_{Mn}, 2 mg/L and MAC_{Fe}, 8 mg/L) [30] in both VSR and GR dessert wines and it is, therefore, assumed that those elements did not represent any threat for the quality of the products and safety of the consumers.

Table 2. Contents of metal nutrients (C_{Na}, C_K, C_{Mg}, C_{Ca}, C_{Mn}, C_{Cu}, C_{Zn}, as mg/L) in Vin Santo (VSR) wines as a function of time (days).

Day	C_{Na}	C_K	C_{Mg}	C_{Ca}	C_{Mn}	C_{Fe}	C_{Cu}	C_{Zn}
58th	20.5 ± 0.6	708 ± 18	175 ± 5	36 ± 3	1.80 ± 0.06	1.30 ± 0.07	2.68 ± 0.09	48.5 ± 4.6
59th	20.1 ± 0.5	607 ± 26	177 ± 5	40 ± 4	1.79 ± 0.03	1.33 ± 0.05	2.67 ± 0.11	49.9 ± 7.9
60th	20.2 ± 0.6	650 ± 18	176 ± 5	40 ± 3	1.78 ± 0.06	1.31 ± 0.07	2.72 ± 0.09	50.8 ± 4.6
61th	19.8 ± 0.3	695 ± 21	178 ± 5	38 ± 5	1.70 ± 0.03	1.28 ± 0.04	2.65 ± 0.13	49.1 ± 7.8
62th	19.5 ± 0.3	689 ± 25	180 ± 6	36 ± 2	1.75 ± 0.06	1.29 ± 0.05	2.71 ± 0.12	47.5 ± 4.7
63th	19.6 ± 0.3	695 ± 27	182 ± 5	36 ± 3	1.82 ± 0.06	1.32 ± 0.06	2.74 ± 0.13	50.5 ± 3.7
64th	19.2 ± 0.3	692 ± 20	175 ± 6	42 ± 5	1.80 ± 0.06	1.31 ± 0.04	2.67 ± 0.14	48.5 ± 2.9
65th	19.4 ± 0.4	688 ± 22	178 ± 6	45 ± 4	1.77 ± 0.07	1.34 ± 0.05	2.63 ± 0.15	48.7 ± 3.0
66th	19.6 ± 0.2	690 ± 16	174 ± 6	45 ± 2	1.79 ± 0.07	1.28 ± 0.04	2.66 ± 0.12	49.9 ± 3.6
67th	20.1 ± 0.6	695 ± 16	165 ± 10	44 ± 4	1.80 ± 0.05	1.30 ± 0.04	2.65 ± 0.16	51.0 ± 2.7
68th	20.6 ± 0.5	694 ± 27	170 ± 9	45 ± 4	1.81 ± 0.04	1.32 ± 0.05	2.65 ± 0.13	50.2 ± 2.9
70th	20.5 ± 0.3	687 ± 33	172 ± 3	44 ± 4	1.83 ± 0.04	1.27 ± 0.06	2.73 ± 0.12	47.3 ± 3.1
80th	20.7 ± 0.4	691 ± 27	173 ± 7	43 ± 3	1.79 ± 0.05	1.29 ± 0.06	2.65 ± 0.12	50.0 ± 2.6
90th	20.4 ± 0.5	694 ± 29	168 ± 5	41 ± 3	1.81 ± 0.06	1.30 ± 0.05	2.66 ± 0.16	47.7 ± 2.0
100th	20.5 ± 0.6	693 ± 26	169 ± 9	41 ± 2	1.78 ± 0.05	1.33 ± 0.06	2.64 ± 0.15	49.6 ± 3.7
110th	19.8 ± 0.4	689 ± 28	179 ± 8	41 ± 2	1.82 ± 0.04	1.27 ± 0.06	2.68 ± 0.13	49.6 ± 3.3
Average *	20.0 ± 0.5	685 ± 24	174 ± 5	41 ± 4	1.79 ± 0.03	1.31 ± 0.03	2.68 ± 0.03	49.3 ± 1.2
9th	35.0 ± 0.5	780 ± 28	155 ± 12	61 ± 5	1.76 ± 0.05	1.36 ± 0.08	2.77 ± 0.15	45.0 ± 2.6
Δ% *	−43	−11	12	−32	2	−4	−3	10

Values (mean ± ESD) obtained from three samples, and three replicates each sample; full experimental data sets are reported in Table S2. * Average values calculated from 58th–110th data; Δ% Relative percentage change = [(Average (58th − 110th) − 9th)/9th] × 100.

Finally, the content of Pb, C_{Pb}, in the two dessert wines were 5.6 ± 0.4 (GR) and 15.7 ± 0.3 μg/L (VSR; Table S3), being in both cases, at least two orders of magnitude lower than the maximum allowable concentration enforced by European and Italian laws (MAC_{Pb}, 150 μg/L) [31]. It was not considered necessary to perform any specific lowering strategy for that metal.

2.2. *Assessing the Effect of Synthetic Hydrogels*

Once the content of each metal was considered constant, the assessments of effects by CMH2 and CMH10 hydrogels (Scheme 1) were performed on GR wines. On soaking the not swelled hydrogels (20 mg hydrogel (CMH2-nSW and CMH10-nSW) in 10 mL of GR wine) no appreciable lowering of metal content within 48 h was observed. In the case the histidine-based hydrogels (CMH2, CMH10) were previously swelled in ultrapure water and then soaked (48 h) into the GR wine (20 mg hydrogel/10 mL GR, 20 ± 1 °C, in the dark, not stirred systems) a lowering effect was revealed (Figure 1 and Table S4a,b). The data showed the following: (i) the effects on C_{Na} were null; (ii) the effects on C_K were small (4 ± 1%); (iii) the lowering effects on C_{Ca} and C_{Zn} were small (*ca* 10 ± 3%); (iv) the highest effects on C_{Mg} were 25 ± 3% and those on C_{Cu} were by 28 ± 5% (by using CMH10 material).

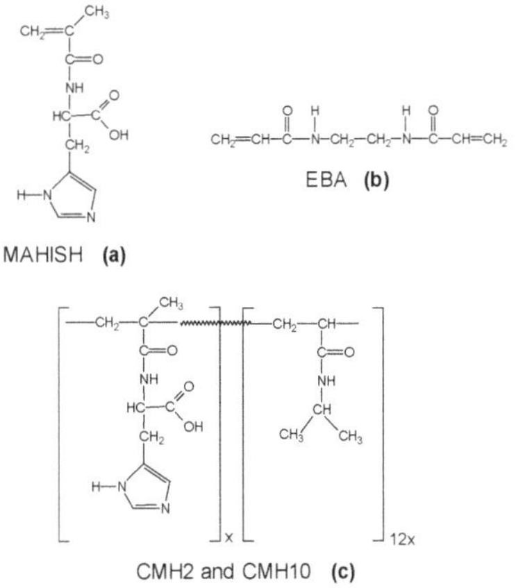

Scheme 1. Structural formula for (**a**) monomer *N*-methacryloyl-L-histidine (MAHISH), (**b**) cross-linker, *N,N'*-ethylene-bis-acrylamide (EBA) and (**c**) general formula for linear polymers CMH2 and CMH10, poly(*N*-methacryloyl-L-histidine-co-*N*-isopropylacrilamide), NIPAAM/MHIST molar ratio 12, cross-linked with 2 and 10 mol% of EBA, respectively.

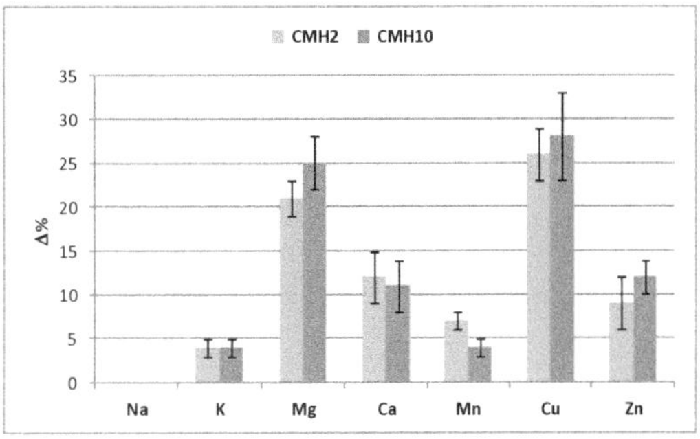

Figure 1. The lowering effects (Δ%) by swelled hydrogels on the content of selected metals in Granello (GR) wine (hydrogel/wine, 20 mg dry matter/10 mL, 2g/L; not stirred, in the dark, 20 ± 1 °C, 48 h). Values (mean ± ESD) obtained from three samples, and three replicates of each sample.

2.3. *Assessing the Effect of Ion Exchange Resins*

The subsequent step was that of determining the effects by solid state commercial resins of the ionic exchange type on samples of dessert wines. The selected commercial resins were Lewatit-207 (L-207) and Lewatit-208 (L-208), consisting of a matrix of polystyrene and divinylbenzene polymers with chelating functional groups based on iminodiacetate [32], able to capture mostly divalent metal cations (Cu, Pb, Ni, Zn, Cd, Fe, Be, Mn, Ca, Mg, Sr and Ba). It has to be noted that similar resins have been used for a long time in the food industry in several countries, including those within the European Union, for the treatments of waters for human uses, as water and fruit juice sweeteners, as acids for food uses, and other similar purposes [3–5,8,33,34].

A preliminary assay was carried out by using 300 mg of resin (L-207 or L-208; immersion beds, previously activated) on 15 mL of VSR samples. The data for the C_{Mn}, C_{Fe}, C_{Cu} and C_{Zn} in VSR wine are reported in Figure 2 (and Table S5a,b). Both the resins had large decreasing effects for C_{Mn} and C_{Fe} (*ca* 80–90%), and significant decreasing effects for C_{Cu} and C_{Zn}, and L-207 had lowering effects larger than L-208 as regards C_{Zn}.

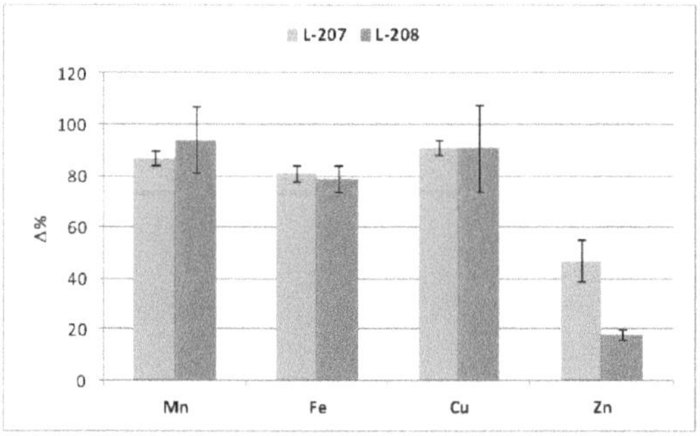

Figure 2. The lowering effects (Δ%) by the resins L-207 and L-208 on the concentrations of micronutrients in VSR wine (resin beds/wine 300 mg/15 mL, 20g/L; not stirred, in the dark, at 20 ± 1 °C, for 48 h). Values (mean ± ESD) obtained from three samples, and three replicates each sample.

2.4. Large Scale Study on L-207 Resin over Increasing Time

A larger scale experiment was carried out by using L-207 activated resin (20 g) for treating 1.00 L of wine sample (both VSR and GR). The data for the effects on the content of metals at different soaking times were recorded (the change of wine volume after each sampling, of 5 mL, was taken into account). The data are reported in Tables 3 and 4 (see also Figure S3).

It is evident that the lowering of C_{Mn}, C_{Cu} and C_{Zn} in GR was very effective. In fact, the C_{Mn} after 48 h treatment (resin bed *ca* 2.0% *w*/*v*, dark) was reduced to 23% of the initial value, and final C_{Mn} was as low as 0.36 ± 0.03 mg/L. The C_{Cu} in the same condition was also much reduced, to 24% of the initial value (0.39 ± 0.01 mg/L). In the case of C_{Zn} the lowering was even larger, the concentration decreased to 8% of the initial value for GR (0.58 ± 0.02 mg/L). The effect on the concentration of alkaline metals was small (C_{Na}, lowering to 87%; C_K, lowering to 85%) and is related to the low charge/surface ratio for the ions. The effects on C_{Mg} and C_{Ca} lowered by 57 and 54% based on untreated GR wine.

Table 3. Content of metal nutrients (C_{Na}, C_K, C_{Mg}, C_{Ca}, C_{Mn}, C_{Cu}, C_{Zn}, as mg/L) in GR wines, as a function of soaking time (up to 48 h) on L-207 resin beds (resin beds/wine 20 g/L, under stirring, in the dark, 20 ± 1 °C).

Time	C_{Na}	C_K	C_{Mg}	C_{Ca}	C_{Mn}	C_{Cu}	C_{Zn}
0	28.3 ± 0.3	754 ± 17	200 ± 6	60 ± 3	1.56 ± 0.02	0.39 ± 0.01	0.58 ± 0.02
1	29.2 ± 2.9	750 ± 26	137 ± 9	43 ± 1	0.63 ± 0.05	0.27 ± 0.03	0.31 ± 0.04
3	27.2 ± 2.9	702 ± 18	124 ± 3	38 ± 3	0.46 ± 0.03	0.20 ± 0.02	0.20 ± 0.02
6	28.8 ± 2.9	672 ± 24	120 ± 3	37 ± 2	0.40 ± 0.05	0.15 ± 0.01	0.12 ± 0.02
18	27.5 ± 3.5	706 ± 17	117 ± 3	34 ± 2	0.37 ± 0.04	0.13 ± 0.01	0.093 ± 0.006
21	25.8 ± 3.4	743 ± 20	116 ± 1	33 ± 3	0.37 ± 0.04	0.12 ± 0.02	0.086 ± 0.006
24	25.8 ± 3.4	678 ± 17	115 ± 3	33 ± 4	0.36 ± 0.02	0.11 ± 0.01	0.081 ± 0.007
30	26.0 ± 3.7	684 ± 16	114 ± 2	33 ± 3	0.36 ± 0.02	0.10 ± 0.02	0.069 ± 0.008
42	24.5 ± 4.1	707 ± 15	114 ± 2	32 ± 4	0.36 ± 0.02	0.098 ± 0.010	0.058 ± 0.005
48	24.9 ± 3.2	643 ± 20	113 ± 2	32 ± 3	0.36 ± 0.03	0.095 ± 0.009	0.047 ± 0.008

Values for t = 0 are those from untreated wines. Values (mean ± ESD) obtained from three samples, and three replicates for each sample; the full experimental data sets are reported in Table S6.

As regards VSR wine, it has to be recalled that C_{Mn}, C_{Fe}, C_{Cu} and C_{Zn} were higher than the corresponding ones in GR. In the case of Mn the difference between the initial contents C_{Mn}, was limited: 1.79 ± 0.03 (VSR) *contra* 1.56 ± 0.02 mg/L (GR). The difference increased for the C_{Cu} and C_{Zn} values. The trend of the lowering ratio versus the charge/surface ratio was almost linear for the micronutrients in VSR, R^2 = 0.895, on passing from Mn to Zn, when the oxidation status for Fe was assumed as +2; whereas the data for Fe^{+3} did not fit the trend, confirming that iron is reasonably in the reduced state. The relative lowering effects on C_{Mn}, C_{Fe}, C_{Cu} and C_{Zn} were down to 25, 15, 9, and 2%, respectively, of the initial values. It was confirmed that macronutrients Na, K, Mg and Ca had almost the same concentrations in VSR with respect to GR wines. The C_{Na}, and C_K were reduced just down to about 95% of the initial values, whereas C_{Mg} and C_{Ca} were reduced down to 61 and 44% of the initial values, by L-207. Interestingly, C_{Zn} in VSR samples averaged 49.3 ± 1.2 mg/L, resulting *ca* 10 times higher than the maximum value allowed by the enforced regulation (MAC_{Zn}, 5 mg/L) [29]. Therefore, the optimized treatments, just above those reported, allowed a lowering of the C_{Zn} down below the MAC_{Zn} value, within *ca* 3 h of immersion (of 20.0 g of L-207 resin per 1.0 L of wine under stirring at 20 °C in the dark). The lowering to below 2 mg/L is reachable after 6 h treatment under stirring at a plateau that had C_{Zn} 1.1 ± 0.1 mg/L.

Table 4. Content of metal nutrients (C_{Na}, C_K, C_{Mg}, C_{Ca}, C_{Mn}, C_{Fe}, C_{Cu}, C_{Zn}, as mg/L) in VSR wines, as a function of soaking time (up to 48 h) on L-207 resin beds (resin beds/wine 20 g/L, under stirring, in the dark, 20 ± 1 °C).

Time	C_{Na}	C_K	C_{Mg}	C_{Ca}	C_{Mn}	C_{Fe}	C_{Cu}	C_{Zn}
0	20.0 ± 0.5	685 ± 24	174 ± 5	41 ± 3	1.79 ± 0.03	1.31 ± 0.03	2.68 ± 0.03	49.3 ± 1.2
1	20.1 ± 2.2	648 ± 20	138 ± 3	23 ± 2	0.68 ± 0.10	0.50 ± 0.07	1.10 ± 0.11	5.24 ± 1.42
3	21.0 ± 1.8	690 ± 18	128 ± 2	18 ± 1	0.53 ± 0.04	0.40 ± 0.05	0.89 ± 0.14	2.07 ± 0.16
6	19.7 ± 2.3	682 ± 18	123 ± 1	17 ± 2	0.47 ± 0.04	0.34 ± 0.03	0.67 ± 0.12	1.26 ± 0.06
18	20.6 ± 2.9	674 ± 19	128 ± 4	17 ± 2	0.47 ± 0.03	0.28 ± 0.02	0.41 ± 0.04	1.21 ± 0.04
21	20.5 ± 2.8	666 ± 17	125 ± 2	18 ± 1	0.47 ± 0.03	0.26 ± 0.02	0.35 ± 0.03	1.06 ± 0.05
24	20.1 ± 2.3	649 ± 22	119 ± 3	18 ± 2	0.45 ± 0.02	0.23 ± 0.03	0.35 ± 0.02	1.34 ± 0.06
30	20.6 ± 2.7	686 ± 20	122 ± 3	17 ± 2	0.44 ± 0.03	0.21 ± 0.03	0.31 ± 0.02	1.00 ± 0.06
42	20.3 ± 2.4	681 ± 19	116 ± 2	18 ± 1	0.47 ± 0.03	0.20 ± 0.03	0.25 ± 0.02	0.94 ± 0.06
48	19.4 ± 3.0	655 ± 26	107 ± 2	18 ± 1	0.45 ± 0.04	0.20 ± 0.02	0.25 ± 0.02	1.06 ± 0.08

Values for t = 0 are those from untreated wines. Values (mean ± ESD) obtained from three samples, and three replicates for each sample; the full experimental data sets are reported in Table S7.

2.5. *Effects of L-207 on Wine Character: pH and Color Index Indicator Parameters*

The values of pH and color index (CI) were selected as indicator parameters for checking the overall macroscopic quality of the wines, during the treatments (Table 5 and Figure S4).

Table 5. Trend for pH and color index (CI; Abs, λ 420 nm) values in GR and VSR wines as a function of soaking time (up to 48 h) on L-207 resin (resin beds/wine 20 g/L, under stirring, dark, 20 ± 1 °C).

Time	GR		VSR	
	pH	CI (Abs Units) *	pH	CI (Abs Units) *
0	3.47 ± 0.03	0.437 ± 0.007	3.45 ± 0.02	1.056 ± 0.014
6	3.32 ± 0.02	0.366 ± 0.012	3.24 ± 0.03	0.912 ± 0.009
18	3.33 ± 0.02	0.372 ± 0.012	3.24 ± 0.02	0.903 ± 0.009
24	3.32 ± 0.03	0.367 ± 0.007	3.23 ± 0.02	0.906 ± 0.008
36	3.31 ± 0.02	0.364 ± 0.006	3.21 ± 0.02	0.901 ± 0.007
48	3.32 ± 0.01	0.366 ± 0.008	3.22 ± 0.02	0.902 ± 0.006

Values for t = 0 are those from untreated wines. Values (mean ± ESD) obtained from three samples, and three replicates each sample; the full experimental data sets are reported in Tables S8 and S9. * Data collected from wine samples treated with regenerated L-207 resin.

The initial pH value for GR wine before treatment with L-207 resin was 3.47 ± 0.03, then a plateau by 3.32 ± 0.02 was reached after 6 h (decreasing by 0.15 units, 4.3%); whereas for VSR the shift was from 3.45 ± 0.03 for untreated wine to a plateau by 3.23 ± 0.02 (decreasing by 0.22 units, 6.4%).

The color index (CI) values for GR wine changed from an initial absorbance value by 0.437 ± 0.007, to a plateau by 0.366 ± 0.012 after 6 h treatment (decreasing by 0.071 Abs units, 16%). As regards VSR, the CI change from the initial value by 1.056 ± 0.014, was stabilized down to a plateau by 0.912 ± 0.009 Abs unit (decreasing by 0.144 Abs unit, 14%, in 6 h treatment).

3. Discussion

3.1. Stabilization of Metals in GR and VSR Dessert Wines

Selected macro- and micro-nutrients were analyzed on GR and VSR wine samples, from the sampling time up to stabilization, that occurred within *ca* 2 months. The changes of the metal contents for the samples taken from the wood casks are due to both thermal and microbiological processes, that occurred during the storage at laboratory conditions (20 ± 1 °C). The samples were not homogeneous, at the very collection time and had evident precipitates/haziness due to the possible presence of yeasts, metal-tartrates, -sulfides, -carbonates, etc. It has to be noticed that the C_{Mn} in high-quality red Chianti wines from the same company, as previously determined from this laboratory, and coming from harvests in 2001–2008, were in the ranges 2.03 ± 0.01–1.26 ± 0.01 mg/L [27]. The previously reported C_{Fe}, C_{Cu} and C_{Zn} for top-quality red wines ranged 4.68 ± 0.05–1.03 ± 0.01 mg/L, 0.20 ± 0.02–trace (<0.100) mg/L, and 1.10 ± 0.04–0.53 ± 0.03 mg/L, respectively [27]. Therefore, a good agreement was detected between present work values for the GR and VRS dessert wines and previously studied red wines, even though some differences are evident, particularly for Cu and Zn (in VSR). The diversity of fermentation, aging and storing procedures, as well as large differences in grape varieties, but similarity about "terroir" played some role.

In conclusion the stabilization, at controlled temperature for both GR and VSR dessert wines had to be allowed, and immediately after that period and after the relevant analytical determinations, proper treatments for decreasing the content of certain metals could be performed. The bottling and commercialization could occur afterwards.

3.2. The Effect of Synthetic Hydrogels

Regarding the novel application of the synthetic hydrogels, it has to be noted that these materials are expensive in terms of working time and cost of synthesis, purification, swelling, and regeneration. Furthermore, the hydrogels had a non-negligible erosion (qualitatively estimated by *ca* 5% each step, even at rest) and are not approved yet for the food and beverage industry to the best of our knowledge. The relative amounts of hydrogels used in these preliminary tests were small (20 mg/10 mL, *ca* 0.2% *w*/*w*). The efficiency for Cu^{2+} decreasing was 26 ± 3 and 28 ± 5% for CMH2 and CMH10 hydrogels, respectively. Therefore, that route does not have to be discarded in principle, but is worthy of further research. The synthetic hydrogels used have never been tested for wine treatments, neither for research or industrial purposes, to our knowledge. It can be noticed that very recently a paper by Friedenberg and co-workers (2018) [11] reported the usage of polyvinylimidazole and polyvinylpyrrolidone copolymers (PVI-PVP) to reduce copper content in white wines. The latter materials have been reported for several years [4] and are actually also accepted by European regulation (since 2009 [15]), but they were different from the "smart" hydrogels presented in this work. These materials have a great application as carriers of organic molecules, metal-complexes, anions and cations in bio-systems [24,35–37]. The study of these materials could be expanded in future investigations and papers, possibly by: (1) using larger amounts of hydrogels, (2) optimizing the nature of the functional active groups, (3) optimizing reticulation percentage, (4) determining erosion effects, (5) determining effects on wine organoleptic properties, (6) measuring the toxicity of hydrogels. Of course, all of that is not possible in this paper. Finally, we wished to recall that this project managed to compare the effects of dry hydrogels and pre-swelled ones, because in case the dry materials would have worked, the potential industrial application could skip that step. Unfortunately, dry materials did not work and only water pre-swelled gels were able to decrease the content of certain metals.

3.3. The Effect of Ion Exchange Resins

The two commercial resins were efficacious, and were easier to manipulate than the synthetic hydrogels. Furthermore, the resin L-207 was considered more suitable than L-208, because of its lowering effect for C_{Zn}. Based on these data, it was decided to extend the analysis of the effects by L-207, at larger scale in VSR and GR dessert wines. The results showed a great quenching for Cu^{2+} and Zn^{2+} that were the main focus of the work particularly in VRS wine. This was reached within a few hours of soaking for the resin under moderate stirring, to guarantee a limited erosion (that was below 2% each treatment) and possibly low changes of general flavors and other tasting properties. In fact, both VSR and GR wines underwent a moderate decrease of pH and color index upon treatments, suggesting that even the tastes of the products and the content of polyphenols and esters were not much altered by the study treatments. Finally, it is interesting to underline that these resins were already approved for the food and beverage industry. If a wine-making company treats *ca* 90 L of VSR, reduces the concentration of Zn from 49.3 ± 1.2 to 1.1 ± 0.1 mg/L, and mixes the new treated wine with *ca* 10 L of untreated wine, 100 L wine at C_{Zn} < 5 mg/L (MAC_{Zn}) can be obtained. This imply that the flavor of the final product might be altered at lesser extent, with respect to the treatment of the full wine amount, and this allows also a lower resin and general handling working cost. That general strategy was considered better than constructing plants with columns packed with the resin. This latter should use a system of pumps and valves, that would add the risk of contamination of the wine through the contact of the product with mechanical devices.

Attention was also devoted to the ecological aspects related to the treatments of dessert wines with L-207 resin, requesting activation and regeneration processes. The waste materials are in fact aqueous acidic solutions, but their acidity and amount are small when compared to the amount of treated wine. Hypothesizing the treatment of 1 hL of wine, the usage of about 5 L of HCl (2M) activating solution, 10 L of water used for the washing, and 2 L of discarded acidified wine, can be predicted. These acidic solutions need to be neutralized by treating them with NaOH (*ca* 0.4 kg), before being discarded as waste (*ca* 15 L of salt solution). Furthermore, the turnover of the resin has to be considered, mostly because of erosion and reduced activity. On assuming an erosion by *ca* 2% each treatment cycle (1 hL), times 10 cycles (10 hL of treated wine), it means that less than 500 g of resin are predicted to be disposed to waste. This corresponds to 2000 bottles (0.5 L each) of the final product, ready for the market. The exhausted resin could be sent back to the resin producer for overall reconditioning.

4. Materials and Methods

4.1. Chemicals and Materials

Standard solution for atomic absorption analyses were purchased from Sigma-Aldrich (Milan, Italy) as TraceCERT single metal 1000 mg/L of Na, K, Ca, Mg, Mn, Fe, Cu, Zn, and Pb solution in 0.2% nitric acid. Suprapur nitric acid (65%), L-histidine, acryloyl chloride, *N*,*N*′-ethylene-bis-acrylamide (EBA), *N*-isopropylacrylamide (NIPAAM) were also purchased from Sigma-Aldrich (Milan, Italy). Ionic exchange resins Lewatit L-207 and L-208 were purchased from Lanxess (Colonia, Germany; Table 6). The two poly(*N*-methacryloyl-co-*N*-isopropylacrylamide) hydrogels (CMH2 and CMH10) were synthesized as previously reported [35–37]. In both cases, the NIPAAm/MHist molar ratio was 12 with different EBA (*N*,*N*′-ethylene-bis-acrylamide) cross-linking contents (2 mol%, CMH2; 10 mol% CMH10). The bidistilled water was produced by an Acquinity P/7 distiller (MembraPure GmbH, Berlin, Germany).

Table 6. General characteristics and selected recommended operating conditions, of the commercial resins, Lewatit-207 (L-207) and Lewatit-208 (L-208) (Lanxess, Germany). Data taken from product information.

	Lewatit-207 (L-207)	Lewatit-208 (L-208)
Appearance	Beige, opaque	Beige, opaque
Matrix	Cross-linked polystyrene	Cross-linked polystyrene
Functional group	Iminodiacetic acid	Iminodiacetic acid
Structure	Macroporous	Macroporous
Ionic form, as shipped	Na^+	Na^+
Particle size	0.61 ± 0.05	0.65 ± 0.05
Particle geometry	Sphere	Sphere
Density	1.10 g/mL	1.16 g/mL
Water retention	55–60%	58–64%
Total capacity (H-form)	min 2.0 eq/L	min 2.5 eq/L
Operating temperature	max 80 °C	max 80 °C
Operating pH range	1.5–9	2–12
Regenerant	HCl 7.5%(*w*/*w*) or H_2SO_4 10% (*w*/*w*)	HCl 10% (*w*/*w*)
Conditioning	NaOH 4% (*w*/*w*)	NaOH 4% (*w*/*w*)
Rinse water requirement	5 times total bead volume	5 times total bead volume

4.2. Dessert Wine Samples

As regards VSR dessert wine, Malvasia grape bunches that consisted of good quality healthy berries had been harvested from 12-year-old vines, late in the proper season *ca* half of September and had a high sugar content, that corresponded to a density of the grape juice by *ca* 1.11 g/mL. After drying (standing suspended in well ventilated rooms for several months) the bunches had been crushed lightly and fermented slowly, and then placed in French oak bariques (by 225 L). A subsequent aging period by five years, allowed the final alcohol content to rise up to 15.5 ± 0.2% *v*/*v* (as measured through the official density method [4], over at least three replicates on separate samples). The collected aged VSR wines from the oak casks, were stored in cork sealed bottles (six), in horizontal position, in the dark, at 20 ± 1 °C. They were then submitted to periodic analysis for the content of selected metals, up to stabilization.

As regards GR dessert wine from top quality bunches of Sauvignon grapes, were harvested the same period of VSR, were stored in a ventilated, dried and dehumidified room for 20 days. Then the berries were cooled down to 12 ± 1 °C and mildly pressed. The must was then fermented in steel containers for 30 days, at temperature by 12 ± 1 °C. Then the sweet wine was stored in cork-sealed bottles for six months, in the cellar. Finally, the sampled bottles (six) were stored as just reported for VSR, and submitted to periodic analysis of the content of selected metals, up to stabilization.

4.3. Color Index (CI) Determination

The absorbances at λ = 420 nm, that is the characteristic wavelength for white wines, were acquired for measuring the color index (CI) parameter, by using the Lambda 10 Perkin Elmer spectrophotometer equipped with a continuous flow quartz cuvette (10 mm optical path length, Perkin Elmer (Milan, Italy). All measurements were carried out in triplicate on three samples as such, without any pre-treatment, at 20 ± 1 °C.

4.4. pH Determination

Determinations for pH values were performed through a Crison pH-meter model 507 equipped with a glass combined electrode (Crison, Barcelona, Spain). The instrument was calibrated by using Crison buffer solutions at pH 7.00 ± 0.01 and 4.00 ± 0.01, before performing each series of measurements. All measurements were carried out in triplicate on three samples as such, without any pre-treatment, at 20 ± 1 °C.

4.5. Metal Analysis via Atomic Absorption Spectrophotometry (AAS)

The analyses for the metals Na, K, Ca, Mg, Cu, Fe, Mn, Zn and Pb were performed through the atomic absorption spectrophotometry technique with thermal activation (flame, FAAS; AAnalyst300 spectrophotometer, Perkin-Elmer, Waltham, MA, USA), and graphite furnace electro-thermal activation (GFAAS; Varian SpectrAA 220Z with Zeeman background correction spectrophotometer; Cernusco sul Naviglio, Milan, Italy). Multi-element hollow cathode lamps (Cu-Fe-Mn-Co-Cr-Ni), (Na-K), and (Ca-Mg-Zn) and mono-element (Pb) hollow cathode lamps were used (all from Perkin-Elmer). For FAAS analyses, the flame was fuelled by air/acetylene mixtures, optimized for each element to reach a better signal/noise ratio and better limits of quantification (LOQ) and limits of detection (LOD). In the case of GFAAS analyses, the graphite tubes were of pyrolytic type with integrated L'vov platform (Varian). The absorbance recorded for each sample was an average of 10 readings (FAAS) or three readings (GFAAS). The standard solutions for calibration purposes were obtained by diluting commercial mother solutions (1000 mg/L), with ultrapure 0.2% HNO_3 (65%, Suprapur) solution. The external calibration method was commonly used in properly selected linearity range for each element. Selected samples were also analyzed via the standard addition method [38] to exclude errors from matrix effects, the differences between the data found by the two calibration methods resulting <5%. External calibrations showing correlation factors $R^2 > 0.990$ were accepted for quantitative analyses (Table S10). Other details relevant to the analytical methods can be found from [27,39]. All the metal determinations were carried out in triplicate, on three bottles for each sample. The LOQ for the FAAS methods for the analyzed elements were Na 0.100, K 0.200, Mg 0.050, Ca 0.300, Mn 0.100, Fe 0.200, Cu 0.100, and Zn 0.050 mg/L, respectively; and for the GFAAS method for Pb LOQ was 1.0 µg/L (Table S10).

The contents of selected metal nutrients (C_{Na}, C_K, C_{Mg}, C_{Ca}, C_{Mn}, C_{Fe}, C_{Cu} and C_{Zn}) were analyzed in GR and VRS dessert wines before any treatment, in the time period 58–110th days from the sampling at the wine-making company, in order to evaluate the stability of the metal parameters. A first analysis was also performed at 7th (GR) or 9th (VSR) day from the sampling, to estimate the initial conditions. After the removal of the original seal, a sampling of 2 mL was performed from each bottle ($n = 3$). The bottles were then properly re-sealed with their own cork, surrounded by parafilm and stored in a dark room at 20 ± 1 °C. The analyses were performed on the samples as such, or after proper dilution (if necessary). The same metals, were then analyzed in wine samples, after treatment with synthetic hydrogels or commercial ionic exchange resins (see below), to assess the lowering metal content ability of the materials.

4.6. Hydrogels

4.6.1. Synthesis

Hydrogels were prepared as copolymers of the commercial *N*-isopropylacrylamide (NIPAAm) and the synthetic monomer *N*-methacryloyl-L-histidine, with a radical polymerization reaction in the presence of the cross-linking agent *N,N'*-ethylene-bis-acrylamide (EBA). The specific procedures were those previously reported [35–37]. The synthesized hydrogels are hereafter named CMH2, and CMH10. The formula for the monomer (*N*-methacryloyl-L-histidine, MAHISH), the cross-linker (*N,N'*-ethylene-bis-acrylamide, EBA), and for linear polymers CMH2 and CMH10 (poly(*N*-methacryloyl-L-histidine)-co-*N*-isopropylacrilamide, NIPAAM/MHIST molar ratio 12, cross-linked with 2 and 10 mol% of EBA, respectively; Scheme 1).

4.6.2. Hydrogel Swelling

Several tests were carried out in order to determine the swelling properties of the hydrogels36-38]. On summarizing, aliquots of 20.0 ± 0.1 mg each, of dry CMH2 and CMH10, were swelled in 10 mL of ultrapure water for 48 h. Subsequently, the swelled material was taken off water, set on Whatmann filter paper to remove excess water, and finally weighed. The swelled gels averaged 65.6 ± 5.0 mg for CMH2 and 60.3 ± 5.0 mg for CMH10, respectively (on three replicates). Then, both dry gels and swelled gels were used for preliminary experiments on evaluating the lowering efficacy for metal contents in GR dessert wine. Each experiment was repeated in triplicate.

4.7. Metal Contents in Dessert Wines after Treatments by Synthetic Hydrogels and Commercial Resins

4.7.1. Treatments by Synthetic Hydrogels

The not swelled (CMH2-nSW and CMH10-nsW) and swelled (CMH2 and CMH10) hydrogels (20 mg, dry weight) were set in 10 mL of GR wine sample and stored at 20 ± 1 °C, in the dark, not stirred, in hermetically closed glass vessels, for 48 h before the analyses. Then, an aliquot (5 mL) was taken off from the batch, possible solid materials were filtered off through Whatmann filtering paper, and the clear solution was analyzed for the contents of metals. Each experiment, from gel soaking through the analysis, was performed in triplicate.

4.7.2. Treatments by Lewatit-207 and Lewatit-208 Resins

In preliminary tests, aliquots (15 mL) of clear stabilized VSR wine, were treated with the resin (300 mg; immersion beds). The resins were previously activated (by treating them with HCl 2M (15 mL for 2 h, under stirring), then filtered and subsequently rinsed twice with ultrapure water (1 h each, under stirring), and finally filtered and brought to dryness (in the air). Then, the beds were added to the wine sample in a sealed glass vessel (50 mL overall capacity, 4 cm diameter) equipped with a screwed plasticized cork to guarantee perfect closure, and maintained at rest (not stirred) for 48 h in order to reach equilibrium (20 ± 1 °C, in the dark). After that, the vessels were open and the surnatants were filtered and finally analyzed. Each experiment was repeated in triplicate.

4.7.3. Large-Scale Treatments by Lewatit-207 Resin

The VSR and GR wines (1.00 L each) were set on beds of the activated resins (20.0 g, previously activated by 100 mL of HCl 2 M), in sealed glass vessels (1.5 L overall capacity, 10.6 cm diameter) equipped with a screwed plasticized cork to guarantee the perfect closure. The system was maintained under stirring through a magnetic bar (5.0 cm length, 0.5 cm diameter, iron bar covered with Teflon film, at 60 rpm; 20 ± 1 °C. in the dark, 48 h). Fractions (5 mL each) were collected after 1, 3, 6, 18, 21, 24, 30, 42 and 48 h. Each experiment was repeated in triplicates.

4.8. Statistical Data Treatment

Six bottles of both VSR and GR were sampled. All the experiments were carried out in triplicate and triplicate analyses were performed for all measurements. Mean values and estimated standard deviations (esd) were calculated by using Microsoft Office Excel 2007, implemented with regression analysis subroutine, and Origin Pro8 SR2, v.0891 (B891).

5. Conclusions

The work showed that the resin L-207 had a significant ability to lowering C_{Zn}, and also C_{Mn}, C_{Cu} and C_{Ca} that were high in a certain year's harvest, at least working at the optimized experimental conditions. The content of Mn might be related to the particular nature of the soil for the study area that had a high content of Mn in that area, as confirmed by results previously found for wines produced by the same company [27]. Regarding C_{Zn}, the high value found from the study samples had a rationale in the usage of zinc-plated racks for grape drying steps. The proposed remedy for the study production, was the treatment with L-207 resin. For future production the substitution of the metal racks with polyethylene racks is strongly advisable. Notwithstanding this, the final product would have to be also treated with L-207 for keeping other metals, like copper, manganese and calcium, under control. Finally, it will be important to test the possible changes on organoleptic properties, by using panels of tasters and sommeliers. The capturing effects on selected metal ions by ion exchange resins and hydrogels on table white wines and dry dessert white wines could also be tested, although the protocol is reasonably exportable and applicable to other kind of white wines.

Supplementary Materials: Supplementary data associated with this article can be found online.

Author Contributions: Conceptualization, G.T., R.C. and C.R.; Formal analysis, G.T., A.P., M.C., G.L. and M.C.; Investigation, C.B. and A.D.; Methodology, G.T., R.C. and A.M.; Validation, G.T. and R.C.; Writing–original draft, G.T., C.B. and R.C.; Writing–review and editing, G.T., A.P., C.B., A.D., M.C., G.L., M.C., R.C., A.M. and C.R.

Funding: This research was funded by Provincia di Siena for the grant "Ricercatori in Azienda 2011" to GT and the APC was funded by financial resources from University of Siena.

Acknowledgments: The authors thank the Barone Ricasoli Agricola SpA for the gift of GR and VSR samples and for useful information relevant to wine-making procedures for the dessert wines. GT gratefully acknowledges the Provincia di Siena for the grant "Ricercatori in Azienda 2011". The Department of Biotechnology Chemistry and Pharmacy (Department of Excellence 2018–2022), University of Siena, is also acknowledged.

Conflicts of Interest: The authors declare that they have no conflicts of interest.

References and Notes

1. Domizio, P.; Lencioni, L. Vin Santo. *Adv. Food Nutr. Res.* **2011**, *63*, 41–100. [PubMed]
2. European Community, Commission Regulation (EC). No. 1512/2005 (15 September 2005) amending Regulation (EC) No. 753/2002 laying down certain rules for applying Council Regulation (EC) No. 1493/1999 as regards the description, designation, presentation and protection of certain wine sector products.
3. Jackson, R.S. *Wine Science, Principles and Applications*, 4th ed.; Academic Press: New York, NY, USA, 2014; pp. 143–306.
4. Zoecklein, B.W.; Fugelsang, K.C.; Gump, B.H.; Nury, F.S. *Wine Analysis and Production*; CBS Publishers and Distributors: New Delhi, India, 1997.
5. Vine, R.P.; Harkness, E.M.; Linton, S.J. *Wine Making from Grape Growing to Marketplace*, 2nd ed.; Springer Science + Business Media, Inc.: New York, NY, USA, 2002; p. 85.
6. Walker, T.; Morris, J.; Threlfall, R.; Main, G. pH modification of Cynthiana wine using cationic exchange. *J. Agric. Food Chem.* **2002**, *50*, 6346–6352. [CrossRef] [PubMed]
7. Bonorden, W.R.; Nagel, C.W.; Powers, J.R. The adjustment of high pH/high titratable acidity wines by ion exchange. *Am. J. Enol. Viticult.* **1986**, *37*, 143–148.
8. Ibeas, V.; Correia, A.C.; Jordão, A.M. Wine tartrate stabilization by different levels of cation exchange resin treatments: Impact on chemical composition, phenolic profile and organoleptic properties of red wines. *Food Res. Int.* **2015**, *69*, 364–372. [CrossRef]
9. Lasanta, C.; Caro, I.; Pérez, L. The influence of cation exchange treatment on the final characteristics of red wines. *Food Chem.* **2013**, *138*, 1072–1078. [CrossRef] [PubMed]
10. Ponce, F.; Mirabal-Gallardo, Y.; Versari, A.; Laurie, V.F. The use of cation exchange resins in wines: Effects on pH, tartrate stability, and metal content. *Cienc. Investig. Agrar.* **2018**, *45*, 82–92. [CrossRef]
11. Friedenberg, D.S.; Manns, D.C.; Perry, D.M.; Mansfield, A.K. Removal of copper from white wine: Imidazole-based polymers are efficient at copper adsorption. *Catal. Discov. Pract.* **2018**. [CrossRef]
12. Mira, H.; Leite, P.; Catarino, S.; da Silva, J.M.R.; Curvelo García, A.S. Metal reduction in wine using PVI-PVP copolmer and its effects on chemical and sensory characters. *VITIS J. Grapevine Res.* **2007**, *46*, 138–147.
13. Benítez, P.; Castro, R.; Barroso, C.G. Removal of iron, copper and manganese from white wines through ion exchange techniques: Effects on their organoleptic characteristics and susceptibility to browning. *Anal. Chim. Acta* **2002**, *458*, 197–202. [CrossRef]
14. Guise, R.; Filipe-Ribeiro, L.; Nascimento, D.; Bessa, O.; Nunes, F.M.; Cosme, F. Comparison between different types of carboxylmethylcellulose and other oenological additives used for white wine tartaric stabilization. *Food Chem.* **2014**, *156*, 250–257. [CrossRef] [PubMed]
15. Commission Regulation (EC) No. 606/2009 (10 July 2009) laying down certain detailed rules for implementing Council Regulation (EC) No. 479/2008 as regards the categories of grapevine products, oenological practices and the applicable restrictions.
16. Bonechi, C.; Lamponi, S.; Donati, A.; Tamasi, G.; Consumi, M.; Leone, G.; Rossi, C.; Magnani, A. Effect of resveratrol on platelet aggregation by fibrinogen protection. *Biophys. Chem.* **2017**, *222*, 41–48. [CrossRef] [PubMed]

17. Tamasi, G.; Mangani, S.; Cini, R. Copper(I)-alkyl sulfide and -cysteine tri-nuclear clusters as models for metallo proteins: A structural density functional analysis. *J. Biomol. Struct. Dyn.* **2012**, *30*, 728–751. [CrossRef] [PubMed]
18. Tamasi, G.; Bonechi, C.; Rossi, C.; Cini, R.; Magnani, A. Simulating the active sites of Copper trafficking proteins. Density Functional Structural and spectroscopy studies on Copper(I) complexes with thiols, carboxylato, amide and phenol ligands. *J. Coord. Chem.* **2016**, *69*, 404–424. [CrossRef]
19. Tamasi, G.; Bonechi, C.; Donati, A.; Leone, G.; Rossi, C.; Cini, R.; Magnani, A. Analytical and structural investigation via infrared spectroscopy and Density Functional Methods of cuprous complexes of the antioxidant tripeptide glutathione (GSH). Synthesis and characterization of a novel CuI-GSH compound. *Inorg. Chim. Acta* **2018**, *470*, 158–171. [CrossRef]
20. Tamasi, G.; Merlino, A.; Scaletti, F.; Heffeter, P.; Legin, A.A.; Jakupec, M.A.; Berger, W.; Messori, L.; Keppler, B.K.; Cini, R. {Ru(CO)x}-core complexes with benzimidazole ligands: Synthesis, X-ray structure and evaluation of anticancer activity in vivo. *Dalton Trans.* **2017**, *46*, 3025–3040. [CrossRef] [PubMed]
21. Pontillo, N.; Ferraro, G.; Messori, L.; Tamasi, G.; Merlino, A. Ru-Based CO releasing molecules with azole ligands: Interaction with proteins and the CO release mechanism disclosed by X-ray crystallography. *Dalton Trans.* **2017**, *46*, 9621–9629. [CrossRef] [PubMed]
22. Tamasi, G.; Carpini, A.; Valensin, D.; Messori, L.; Pratesi, A.; Scaletti, F.; Jakupec, M.; Keppler, B.; Cini, R. {Ru(CO)x}-core complexes with selected azoles: Synthesis, X-ray structure, spectroscopy, DFT analysis and evaluation of cytotoxic activity against human cancer cells. *Polyhedron* **2014**, *81*, 227–237. [CrossRef]
23. Casolaro, M.; Bignotti, F.; Ferruti, P. Amphoteric poly(amido[ndash]amine) polymers containing the ethylenediamine-*N*,*N*′-diaceticacid moiety Stability of copper(II) and calcium(II) chelates. *J. Chem. Soc. Faraday Trans.* **1997**, *93*, 3663–3668. [CrossRef]
24. Tamasi, G.; Casolaro, M.; Magnani, A.; Sega, A.; Chiasserini, L.; Messori, L.; Gabbiani, C.; Valiahdi, S.M.; Jakupec, M.A.; Keppler, B.K.; et al. New platinum–*oxicam* complexes as anti-cancer drugs. Synthesis, characterization, release studies from smart hydrogels, evaluation of reactivity with selected proteins and cytotoxic activity in vitro. *J. Inorg. Biochem.* **2010**, *104*, 799–814. [CrossRef] [PubMed]
25. Bonechi, C.; Donati, A.; Tamasi, G.; Leone, G.; Consumi, M.; Rossi, C.; Lamponi, S.; Magnani, A. Protective effect of quercetin and rutin encapsulated liposomes on induced oxidative stress. *Biophys. Chem.* **2018**, *233*, 55–63. [CrossRef] [PubMed]
26. Leone, G.; Consumi, M.; Pepi, S.; Lamponi, S.; Bonechi, C.; Tamasi, G.; Donati, A.; Rossi, C.; Magnani, A. New formulations to enhance lovastatin release from Red Yeast Rice (RYR). *J. Drug Deliv. Sci. Technol.* **2016**, *36*, 110–119. [CrossRef]
27. Tamasi, G.; Owens, N.F.; Cascella, F.; Cerqua, M.; Cini, R. A Case Study of Selected Volatile Phenols from *Brettanomyces* and Micronutrients Mn, Fe, Cu, Zn in Chianti Red Wines. *J. Food Res.* **2013**, *2*, 31–40. [CrossRef]
28. Commission Regulation (EC) No. 1622/2000 (24 July 2000) Laying down Certain Detailed Rules for Implementing Regulation (EC) No. 1493/1999 on the Common Organisation of the Market in Wine and Establishing a Community Code of Oenological Practices and Processes. 2001. Available online: https://publications.europa.eu/en/publication-detail/-/publication/6a12217e-217d-4e9e-a7ed-9522598672fa/language-en (accessed on 13 November 2018).
29. DM 10 August 2017, Italian Ministry of Agricultural and Forestry Policies. Limits of Some Components in Wines. Application of Art. 25 of Law 12 December 2016, n. 238; Published on GU n. 201, 29 August 2017. Available online: https://www.politicheagricole.it/flex/cm/pages/ServeBLOB.php/L/IT/IDPagina/12019 (accessed on 13 November 2018).
30. Australian Government, Wine Australia: Manganese Levels in China. Available online: https://www.wineaustralia.com/getmedia/c036a4f8-5a36-459b-be60-2f0a36adb277/Manganese-Levels-in-China.pdf (accessed on 13 November 2018).
31. Commission Regulation (EC) No. 1881/2006 (19 December 2006) Setting Maximum Levels for Certain Contaminants in Foodstuffs. Available online: https://eur-lex.europa.eu/legal-content/EN/ALL/?uri=celex%3A32006R1881 (accessed on 13 November 2018).
32. LANXESS Italy Energizing Chemistry, Technical Product Information. Available online: https://www.lenntech.com/Data-sheets/Lewatit-TP-207-L.pdfandhttps://www.lenntech.com/Data-sheets/Lewatit-TP-208-L.pdf (accessed on 13 November 2018).

33. Lasanta, C.; Caro, I.; Pérez, L. Theoretical model for ion exchange of iron(III) in chelating resins: Application to metal ion removal from wine. *Chem. Eng. Sci.* **2005**, *60*, 3477–3486. [CrossRef]
34. Alberti, G.; Pesavento, M.; Biesuz, R. A chelating resin as a probe for the copper(II) distribution in grape wines. *Polymers* **2007**, *67*, 1083–1093. [CrossRef]
35. Casolaro, M.; Ito, Y.; Ishii, T.; Bottari, S.; Samperi, F.; Mendichi, R. Stimuli-responsive poly(ampholyte)s containing L-histidine residues: Synthesis and protonation thermodynamics of methacrylic polymers in the free and in the cross-linked gel forms. *eXPRESS Polym. Lett.* **2008**, *2*, 165–183. [CrossRef]
36. Casolaro, M.; Bottari, S.; Cappelli, A.; Mendichi, R.; Ito, Y. Vinyl polymers based on L-histidine residues. Part 1. The thermodynamics of poly(ampholyte)s in the free and in the cross-linked gel form. *Biomacromolecules* **2004**, *5*, 1325–1332. [CrossRef] [PubMed]
37. Casolaro, M.; Bottari, S.; Ito, Y. Vinyl polymers based on L-histidine residues. Part 2. Swelling and electric behavior of smart poly(ampholyte) hydrogels for biomedical applications. *Biomacromolecules* **2006**, *7*, 1439–1448. [CrossRef] [PubMed]
38. Steliopoulos, P. Extension of the standard addition method by blank addition. *MethodsX* **2015**, *2*, 353–359. [CrossRef] [PubMed]
39. Tamasi, G.; Cambi, M.; Gaggelli, N.; Autino, A.; Cresti, M.; Cini, R. The content of selected minerals and vitamin C for potatoes (*Solanum tuberosum* L.) from the high Tiber Valley area, Southeast Tuscany. *J. Food Compos. Anal.* **2015**, *41*, 157–164. [CrossRef]

Sample Availability: Samples of the compounds are not available from the authors.

molecules

MDPI

Article

A Human Health Risk Assessment of Trace Elements Present in Chinese Wine

Zhi-Hao Deng [1,2], Ang Zhang [2,*], Zhi-Wei Yang [2], Ya-Li Zhong [2], Jian Mu [2], Fei Wang [2], Ya-Xin Liu [2], Jin-Jie Zhang [2] and Yu-Lin Fang [1,*]

1 College of Enology, Northwest A&F University, Yangling 712100, China; 13389254907@163.com
2 Inspection and Quarantine Technology Center of Qinhuangdao Entry-Exit Inspection and Quarantine Bureau, Qinhuangdao 066004, China; muyishixinrenwei@163.com (Z.-W.Y.); zhongyali@163.com (Y.-L.Z.); mujian@163.com (J.M.); wangfei@163.com (F.W.); liuyaxin@163.com (Y.-X.L.); zhangjinjie@163.com (J.-J.Z.)
* Correspondence: zhanganggrape@hotmail.com (A.Z.); fangyulin@nwsuaf.edu.cn (Y.-L.F.); Tel.: +86-335-5308704 (A.Z.); +86-29-87091874 (Y.-L.F.)

Academic Editors: Rosa Perestrelo and José Sousa Câmara
Received: 9 December 2018; Accepted: 7 January 2019; Published: 11 January 2019

Abstract: The concentrations of trace elements in wines and health risk assessment via wine consumption were investigated in 315 wines. Samples were collected from eight major wine-producing regions in China. The concentrations of twelve trace elements were determined by inductively coupled plasma mass spectrometry (ICP-MS) and Duncan's multiple range test was applied to analyze significant variations ($p < 0.05$) of trace elements in different regions. Based on a 60 kg adult drinker consuming 200 mL of wine per day, the estimated daily intake (EDI) of each element from wines was far below the provisional tolerable daily intake (PTDI). Health risk assessment indicated the ingestion influence of individual elements and combined elements through this Chinese wine daily intake did not constitute a health hazard to people. However, Cr and Mn were the potential contaminants of higher health risk in Chinese wines. The cumulative impact of wine consumption on trace elements intake in the daily diet of drinkers should not be ignored due to the presence of other intake pathways.

Keywords: trace elements; health risk assessment; estimated daily intake; ICP-MS; Chinese wine

1. Introduction

In recent years, wine has become increasingly popular in China, and the production of Chinese wine gradually cannot be underestimated worldwide. Compared with the world wine industry, the Chinese wine industry started late but developed rapidly. In 2016, the production of Chinese wine was 1.14 million liters, ranking sixth in the world [1]. On the other hand, the area of Chinese vineyards is extremely large, totaling 864 thousand hectares in 2016 to rank second in the world [2], so Chinese wines are widely distributed in China. The main areas producing Chinese wine are the eastern foothills of Helan Mountain (HM), the Hexi Corridor (HC), Xinjiang (XJ), the ring around the Bohai Gulf (BG), the Loess Plateau (LP), Southwest Highland (SWH), the Northeast (NE) and Yanhuai Valley (YV). The Helan Mountain region is the most important for the Chinese wine industry and is also the major source of high-quality wine in China. Due to its latitude similarity with Bordeaux, it is also called the Bordeaux of China. The Hexi Corridor region is located in Gansu Province included in the Silk Road Economic Belt proposed by China. The environment here is similar to a cool semidesert that is beneficial to the development of grape aromas responsible for wine flavors attributes. The Xinjiang region is not only the largest Chinese wine-producing area, but the grapes also have high sugar accumulation because winemakers delay harvesting to ensure a fully development of aromas and accompanied by dry climate, so the alcohol content of wine is generally high. The Bohai Gulf region is the earliest wine-producing area in China, and it is the birthplace of many old wine enterprises and has a temperate monsoon climate. The Loess Plateau region is located in the middle part of China, with a continental monsoon climate, and has a long history of viticulture. The Southwest Highland region lies in the Hengduan Mountain area of southwestern China, the climate is similar to an alpine valley with the complex landforms. Because of this region's high accumulated temperature and sufficient light, it is also suitable for grape growth. The Northeast region is situated in Northeast China, where latitude is higher and climate is colder. The temperature here is commonly unfit for most grapes except the amur grape, and thus, the wine produced here has unique characteristics such as deep color and the high concentration of minerals and tannins. The Yanhuai Valley region is a burgeoning wine-producing region in China mainly located in the Huaizhuo basin of northwest Beijing, the unique valley ecological climate here is a paradise for grape growth.

Although wine has been increasingly proven to be beneficial to human health due to the inclusion of a variety of nutrients such as amino acids, vitamins, mineral elements, and polyphenols [3–7], no research has encouraged people to drink wine. In addition to restrictions on alcohol, there is a lack of comprehensive assessment of wine safety risks. Food safety has always been a focus of public attention worldwide, particularly in China. Recently, related issues have increasingly been focalized on government and public opinion. With the destruction and pollution of the environment, some of the pollutants have indirectly emerged in our food. The Joint FAO/WHO Expert Committee on Food Additives (JECFA) reports the tolerance limits for human intake of some pollutants. Arsenic is a trace element and also a high-risk carcinogen with a low lethal dose to humans [8], and the provisional tolerable weekly intake (PTWI) of arsenic is 15.0 μg/kg body weight [9]. Lead is a common toxic metal element that affects the hematological, nervous and reproductive systems of humans and causes pathological changes in organs, leading to a decline in the intelligence quotient (IQ) for children [10]. Cadmium is a nonessential element for the human body but is toxic to the kidneys, bones and cardiovascular system [11], and the provisional tolerable monthly intake (PTMI) of cadmium is 25.0 μg/kg body weight [12]. Chromium is an essential element of the human body, nevertheless, it also harms the human body depending on its valence state. Trivalent chromium is a beneficial essential element, but the toxicity of hexavalent chromium is serious for the human body [13]. Nickel is not an essential trace element, it can cause a variety adverse effects of pulmonary, carbonyl nickel has acute toxicity and carcinogenicity [14], and the provisional tolerable daily intake (PTDI) of nickel based on lowest observed adverse effect level (LOAEL) is 12.0 μg/kg body weight [15]. The above contaminants, especially trace elements, have been detected in previous reports from wine [16]. Although the poisoning caused by drinking wine has not been reported directly, these contaminants are still destructive to the human body due to their chronic and persistent effects. In addition, some beneficial elements in wine such as manganese, cobalt, copper, zinc, molybdenum, aluminum and selenium can also cause adverse reactions if their intake exceeds a specific health indicator. For instance, molybdenum is an indispensable trace element for organisms, but the high concentration of molybdenum intake negatively affects semen quality [17]. The provisional maximum tolerable daily intake of zinc is 1.00 mg/kg body weight [18], and the PTWI of aluminum is 1000 μg/kg body weight [19]. The upper tolerable limits of daily intake for manganese and selenium are 11.0 mg/day [20] and 400 μg/day [21] by a 60 kg adult, respectively. Because compared with other routes, including inhalation and dermal contact, food and drink consumption have been suggested to be a major source of human exposure of trace elements [22,23], we must consider the potential impact of these trace elements from wine on the daily diet of drinkers.

Nevertheless, publications on the concentrations of some adverse trace elements in Chinese wines and health risk assessments of the dietary ingestion of these elements from wine are scant. On the other hand, several agencies and organizations such as the US Environmental Protection Agency (US EPA) and JECFA have provided guidelines on the intake of trace elements by humans. Hence, it is necessary to estimate the health risk of Chinese wine from different wine-producing regions. The main objectives of the present study were: (1) to establish and validate an analytical method for the determination of trace elements in wines by inductively coupled plasma mass spectrometry (ICP-MS) after microwave-assisted digestion; (2) to determine the concentrations of aluminum, arsenic, cadmium, chromium, cobalt, copper, lead, manganese, molybdenum, nickel, selenium and zinc in wines produced from eight different regions of China; (3) to compare the determination levels of different regions with the results reported in previous literature and to analyze the distribution characteristics of trace elements in Chinese wines; and (4) to estimate the dietary intake rates of some elements in Chinese wines through the daily consumption of wine and the health risk of trace elements intake at these rates for Chinese wine.

2. Results and Discussion Sections in Wrong Order–Experimental is Last–Renumber Things Affected

2.1. Method Verification Results

The validation results of the ICP-MS method established in this study for trace elements in wine were shown in the Table 1. The correlation coefficient of calibration curves established by mixing standard solutions with different gradients were 0.983–0.999. According to the IUPAC method, continuous parallel determination of 2% HNO_3 blank solution for 11 times, the limit of detection (LOD) of each element was obtained by the 3-fold standard deviation converting to concentration by calibration curves, and the limit of quantitation (LOQ) of each element was obtained by the 10-fold standard deviation converting to concentration by calibration curves. It could be seen that the range of all elements LOD was 0.024–2.62 μg/L, and the range of all elements LOQ was 0.080–8.06 μg/L. Then, the range of precision for all elements obtained by the relative standard deviations of the above 11 times determined reagent blanks was 0.602–4.81%. For the spiked samples of different concentration levels, the range of spike recovery was 81–137%. The determination results of certificated reference materials (GBW 10010) were all within the error range of the certificate values except for those not detected. It indicated for these that the established method was accurate, effective and suitable for the determination of trace elements in wine.

Table 1. The method verification results for the determination of trace elements in wine by ICP-MS.

Element	Slope	Correlation Coefficient (R^2)	Linear Range (μg/L)	LOD (μg/L)	LOQ (μg/L)	Precision (%, n = 11)	Added (μg/L)	Recovery (%)	GBW 10010 (mean ± SD, μg/L)
^{52}Cr	0.011	0.997	0.250–50.0	0.525	1.73	1.20	50.0	84	ND
							100	85	
							150	91	
^{55}Mn	0.016	0.999	0.250–50.0	0.111	0.366	0.602	500	135	17.1 ± 0.724
							1500	137	
							3000	135	
^{59}Co	0.013	0.999	0.250–50.0	0.024	0.080	3.31	1.00	100	ND
							5.00	90	
							10.0	89	
^{60}Ni	0.003	0.999	0.250–50.0	0.157	0.517	2.60	50.0	83	0.284 * ± 0.046
							100	85	
							150	87	
^{63}Cu	0.006	0.999	0.250–50.0	0.060	0.196	2.93	500	91	4.79 ± 0.289
							1500	122	
							3000	115	
^{66}Zn	0.002	0.999	0.250–50.0	0.780	2.58	1.52	500	93	24.3 ± 1.65
							1500	93	
							3000	93	
^{75}As	0.002	0.999	0.250–50.0	0.167	0.552	0.904	1.00	123	ND
							5.00	118	
							10.0	117	
^{98}Mo	0.006	0.999	0.250–50.0	0.025	0.081	1.61	1.00	108	0.515 ± 0.057
							5.00	95	
							10.0	94	
^{111}Cd	0.003	0.999	0.250–50.0	0.038	0.124	1.02	0.050	135	0.089 * ± 0.009
							0.100	108	
							0.500	99	
^{208}Pb	0.021	0.999	0.250–50.0	0.052	0.171	3.20	1.00	91	0.096 * ± 0.032
							5.00	81	
							10.0	86	
^{27}Al	0.008	0.983	10.0–250	1.31	4.33	4.81	500	107	363 ± 21.4
							1500	100	
							3000	90	
^{77}Se	0.0002	0.997	0.250–50.0	2.62	8.06	4.53	1.00	111	ND
							5.00	124	
							10.0	122	

Notes: * The value was below to LOQ but higher than LOD as the estimated result. The results larger than the linear range were calculated after dilution. ND = not detected. SD = standard deviation.

2.2. Concentrations of Trace Elements in Wine

The concentration range and distribution of twelve elements in 315 wines from eight different regions determined using ICP-MS were shown in boxplot (Figure S1 to Figure S12). As for all wines, the element with the highest average concentration was Mn (3101 μg/L), the element with the lowest average concentration was Cd (0.568 μg/L), and other elements concentrations decreased in the order of Al > Zn > Cu > Cr > Ni > Pb > Se > As > Co > Mo. Irrespective of the origin of the wines, the concentration ranges (in μg/L) of the twelve elements analyzed in all wines were as follows: Cr (76–337), Mn (863–16026), Co (1.15–15.3), Ni (9.67–189), Cu (13.7–543), Zn (77–1670), As (1.54–34.9), Mo (0.390–9.62), Cd (0.076–2.36), Pb (2.06–44.8), Al (90–3202) and Se (6.06–15.1). There were some outliers in the determination results of each element except Se. The structures of the data for the concentrations of each element in study regions were different.

An investigation of the concentrations of trace elements in wine from different regions found that each element was present in different concentrations and distributions in different regions, as shown in Table 2. The measured concentration distribution of Cr showed that all regions had basically the same concentration except the Northeast region, and its average concentration (211 μg/L) was significantly higher than those of the others. The Ni concentration was significantly higher in the Northeast (average 81 μg/L) and Bohai Gulf (average 73 μg/L) regions than in the other regions. The concentration of As was the highest in the Northeast (average 12.2 μg/L) region, which was significantly higher than the concentration in the Xinjiang (average 6.21 μg/L) and Bohai Gulf (average 6.70 μg/L) regions, followed by the Helan Mountain (average 5.94 μg/L), Hexi Corridor (average 5.86 μg/L), Loess Plateau (average 4.02 μg/L), Yanhuai Valley (average 4.43 μg/L) and Southwest Highland (average 4.33 μg/L) regions, whose As concentrations were not significantly different. The same concentration levels of Cd were found in the Xinjiang (average 0.205 μg/L), Helan Mountain (average 0.334 μg/L), Hexi Corridor (average 0.354 μg/L) and Yanhuai Valley (average 0.285 μg/L) regions, and they were the lowest concentrations of Cd among the tested regions, while the highest concentration level was in the Northeast (average 1.35 μg/L) region. The wine of the Xinjiang region had the lowest Pb concentration (average 6.86 μg/L) of all wines, and the wines of Helan Mountain, Hexi Corridor, Yanhuai Valley and Southwest Highland regions had the same Pb concentration level which was higher than Xinjiang region. The Pb concentrations of wine in Northeast, Loess Plateau and Bohai Gulf regions were three higher levels which order was Loess Plateau (average 17.7 μg/L) < Northeast (average 20.6 μg/L) < Bohai Gulf (average 25.2 μg/L). The wine Mn concentrations of the Xinjiang, Helan Mountain, Hexi Corridor, Loess Plateau, Yanhuai Valley and Southwest Highland regions ranged from 863–4780 μg/L, and they were significantly lower there than in the Northeast (average 7807 μg/L) and Bohai Gulf (average 6234 μg/L) regions. The Co concentrations of wine barely did not differ between regions, according to an ANOVA, the Co concentration levels of these regions was ordered Northeast > Loess Plateau and Bohai Gulf > Hexi Corridor > Xinjiang, Helan Mountain and Yanhuai Valley > Southwest Highland. The highest Cu concentration of wine was in the Xinjiang region (average 220 μg/L) but the lowest level was in Yanhuai Valley (average 94 μg/L) and Bohai Gulf (average 97 μg/L) regions. The Zn concentration was divided into four levels: the Northeast region had the highest level (average 724 μg/L), the Bohai Gulf (average 597 μg/L) and Loess Plateau (average 534 μg/L) regions had the second highest level, which was followed by the Helan Mountain (average 378 μg/L), Hexi Corridor (average 394 μg/L), Yanhuai Valley (average 424 μg/L) and Southwest Highland (average 488 μg/L) regions, the last one was in Xinjiang (average 288 μg/L) region. The concentration of Mo was lower in wine in general, and the wine Mo concentration in the Xinjiang region (average 3.43 μg/L) was significantly higher than that in the other regions, moreover, these other regions were approximately in same. The lowest Al concentration was in the Southwest Highland (average 553 μg/L) region, while the highest levels were in the Northeast (average 1058 μg/L) and Loess Plateau (average 1055 μg/L) regions, in addition, the other regions were in same level. In general, the overall Se concentrations of the Northeast (average 12.7 μg/L) region was higher, and that of the other regions were similar to each other.

Table 2. Concentrations (mean ± SD, μg/L) of trace elements in 315 wines of different regions.

Element	NE (n = 30)	XJ (n = 50)	HM (n = 60)	HC (n = 30)	LP (n = 20)	YV (n = 35)	BG (n = 40)	SWH (n = 50)
Cr	211 ± 94d	138 ± 27.1abc	131 ± 28.8abc	140 ± 26.0bc	133 ± 25.5abc	123 ± 22.7ab	151 ± 24.1c	118 ± 33.3a
Mn	7807 ± 3786c	1828 ± 471a	1868 ± 491a	1964 ± 500a	2228 ± 431a	2060 ± 612a	6234 ± 3116b	2283 ± 751a
Co	8.01 ± 2.80d	3.29 ± 1.22ab	3.55 ± 1.22ab	4.04 ± 2.03b	5.05 ± 2.33c	3.43 ± 0.962ab	5.87 ± 2.47c	2.96 ± 1.62a
Ni	81 ± 43.5b	26.3 ± 15.1a	27.5 ± 12.6a	28.6 ± 9.70a	29.7 ± 10.9a	22.6 ± 8.08a	73 ± 38.7b	23.3 ± 10.3a
Cu	170 ± 78d	220 ± 126e	159 ± 69cd	112 ± 37.9ab	150 ± 95bcd	94 ± 59a	97 ± 65a	120 ± 63abc
Zn	724 ± 347e	288 ± 114a	378 ± 131ab	394 ± 168b	534 ± 243cd	424 ± 135b	597 ± 203d	448 ± 239bc
As	12.2 ± 9.20d	6.21 ± 2.19bc	5.94 ± 2.51abc	5.86 ± 2.09abc	4.02 ± 1.74a	4.43 ± 1.27ab	6.70 ± 6.01c	4.33 ± 2.72ab
Mo	2.13 ± 1.12b	3.43 ± 1.84c	1.91 ± 1.10ab	2.12 ± 1.15b	1.92 ± 1.69ab	1.52 ± 1.39ab	1.36 ± 0.578a	1.45 ± 0.990a
Cd	1.35 ± 0.434e	0.205 ± 0.099a	0.334 ± 0.209a	0.354 ± 0.287ab	0.516 ± 0.221bc	0.285 ± 0.147a	1.11 ± 0.535d	0.663 ± 0.535c
Pb	20.6 ± 7.39d	6.86 ± 1.78a	11.4 ± 4.08b	11.0 ± 4.94b	17.7 ± 6.30c	13.5 ± 3.76b	25.2 ± 9.82e	13.6 ± 7.71b
Al	1058 ± 552d	664 ± 403ab	820 ± 377bc	867 ± 658bcd	1055 ± 592d	758 ± 222ab	990 ± 293cd	553 ± 484a
Se	12.7 ± 0.602e	10.1 ± 1.75d	9.40 ± 1.24c	9.75 ± 1.76cd	8.48 ± 1.18a	8.55 ± 1.21ab	9.15 ± 1.22bc	8.36 ± 1.21a

Notes: NE: Northeast; XJ: Xinjiang; HM: Helan Mountain; HC: Hexi Corridor; LP: Loess Plateau; YV: Yanhuai Valley; BG: Bohai Gulf; SWH: Southwest Highland. All results larger than the linear range were calculated after diluting different multiples. The lowercase letters a–e indicate significant variation at the $p < 0.05$ level. SD = standard deviation.

For the determination of these elements in wine, there have been some previous studies that could be used for comparison. Fiket et al. [24] reported that the concentration ranges of Cr, Ni, As and Cd in wines from eastern Croatia were 6.50–31.1 μg/L, 15.3–50.0 μg/L, 0.690–19.1 μg/L and 0.175–1.88 μg/L, respectively. Sperkova and Suchanek [25] showed that the Cr, Ni, As and Pb in Bohemian wines were 18.0–32.0 μg/L, 15.0–53 μg/L, 1.50–6.70 μg/L and 11.0–48.0 μg/L, respectively. The concentration ranges of Cr and As were 20.0–50.0 μg/L and 0.040–0.800 μg/L, respectively, which were studied in Nebbiolo-based wines by Marengo and Aceto [26]. Gremaud et al. [27] quantified elements such as Mn, Zn and Al in their research, and the concentrations of Mn, Zn and Al ranged from 270–600 μg/L, from 340–1140 μg/L and from 180–1100 μg/L, respectively. Geana et al. [28] analyzed the geographical origins of Mn, Co, Cu and Zn in Romanian wines, and the average concentrations of Mn, Co, Cu and Zn were 806 μg/L, 4.35 μg/L, 501 μg/L and 434 μg/L, respectively. Alkış et al. [29] investigated the concentrations of Mn (average 399 μg/L), Co (average 3.37 μg/L), Cu (average 145 μg/L) and Zn (average 1244 μg/L) in Turkish wines. According to comparison, the concentrations of As, Cd and Co in most studies were similar to those of this study, but the concentrations of Mn and Cr in most studies were lower than this study. On the other hand, the highest concentration of As in this study (34.9 μg/L) was far lower than the International Organization of Vine and Wine (OIV) limit of arsenic in wine (200 μg/L) [30]. The highest concentration of Pb in this study was 44.8 μg/L, which was lower than the OIV limit on lead content in wine (150 μg/L) [30], and it was also lower than the People's Republic of China national standard for lead content in wine (0.200 mg/kg) [31]. In addition, the OIV issued the highest limit on Cu in wine (1000 μg/L) [30], and the highest Cu concentration (543 μg/L) in this study was lower than this limit. Similarly, the highest concentration of Zn (1170 μg/L) in all the wines was far less than the OIV limit of Zn in wine (5000 μg/L) [30].

2.3. Analysis of the Characteristics and Source of Trace Elements in Chinese Wine

The eight regions which wines were selected in this study were the main wine-producing regions in China, so the wines in this study essentially represented all Chinese wine. In summary, the different concentrations of each element in the wine of different regions reflected the diversities of environmental condition and element distribution in these regions. The element concentrations of wine of the Northeast region were relatively high. The main reason for these high levels was that the wines selected in the Northeast region contained very sweet wine such as ice wine. The concentrations of all the components in these wines were increased by concentrating the wines. On the other hand, the Northeast area was an old industrial base in China with abundant mineral resources. Because of the development of industry, the soil in these areas such as the cities of Tonghua, Linjiang and Benxi was rich in various trace elements. The grapes grown here and the wine made with these grapes thus contained greater amounts of trace elements or might be contaminated. The concentration distributions of each element except Cu in the wines of the Helan Mountain and Hexi Corridor regions are very close. First, the geographical position of eastern foothills of Helan Mountain and Hexi Corridor were close, so the differences in the concentration levels of the various elements due to geographical origin were very small. Then, the concentration of most elements in these two regions indicated that their wines were less polluted by trace elements than the wines of other regions. However, their difference in Cu content might be due to the use of different grape cultivation methods such as the use of Bordeaux. Because the main component of Bordeaux solution was basic cupric sulphate, and the demand of Bordeaux solution for vineyards varied from region to region, so the different amount of Bordeaux solution used in different regions led to the difference of copper concentration level. The high concentration levels of elements in the wine from the ring around the Bohai Gulf might indicate that this region was polluted with high concentrations of trace elements, especially Mn, Ni, Cd and Pb. The concentration levels of the most elements in wine were the lowest in the Yanhuai Valley comparing for all the regions. The Cu and Mo concentration levels of wine were significantly higher in Xinjiang region than in the wine of other regions, which was also related to the distribution of local elements, and perhaps they could be the characteristic elements of this region. Finally, the elements in the wines of the Loess Plateau and Southwest Highland did not have a particularly prominent parts in this study. A comparison of the characteristics of the different elements in each region could provide a general description of trace element conditions in Chinese wine.

2.4. Estimated Daily Intake of Trace Elements through the Consumption of Wine and Its Health Risk Assessment

To evaluate potential hazards resulting from long-term daily consumption of wine containing these measured elements, we referenced the concept of an estimated daily intake (EDI) for trace elements from wine. As there were no related data about the wine daily consumption rate of drinkers in China, and the target of our assessment was adult drinkers, we assumed that an adult drinker had a daily drinking volume of 200 mL (approximately a glass of wine), which was also similar to the wine consumption rate of 195 g/day used in other assessment study [32]. On the other hand, Chinese drinking habits were based on cups, so the volume of 200 mL wine consumption could also make people more specific understanding about the amounts assessed. Then, because the smaller the body weight, the larger the EDI calculated, we selected 60 kg as the adult drinker average weight for calculations so that the assessment results could be applied to more people. The measured concentration of each element was divided into two parts to evaluate: the used average concentration (mean) represented the general intake, and the used 95% confidence interval upper limit (P95) of the average concentration represented the possible highest intake. Based on the JECFA report of the tolerance limits and perniciousness of pollutants intake, we assessed the EDI of Cr, Mn, Ni, Zn, As, Mo, Cd, Pb, Al and Se from wine of the above eight regions, and the results were shown in Table 3. It could be seen that the EDI (including the P95) of each element in all regions was far lower than the related PTDI. The average EDI of Cr, Mn, Ni, Zn, As, Mo, Cd, Pb, Al and Se for all wines were 0.467, 10.3, 0.124, 1.51, 0.020, 0.007, 0.002, 0.047, 2.70 and 0.032 μg/kg bw/day, respectively. The EDI also basically represented the daily intake per unit weight of these elements through the drinking of 200 mL Chinese wine. For each region, the EDI level was consistent with the concentration level of element. On the other hand, Cr, Mo and Pb did not have clear criteria for the PTDI because of their characteristics. The hexavalent form of chromium was difficult to analyze separately, as chromium (VI) was reduced to chromium (III) in the stomach and gastrointestinal tract [33]. Thus, there were no adequate toxicity studies available to provide a basis for no observed adverse effect level (NOAEL). Molybdenum was ordinarily considered an essential element with an estimated daily requirement of 0.100–0.300 mg for adults [34]. These values were much higher than the EDI obtained in this study, indicating that the intake of Mo from wine hardly affected the intake of Mo in the daily dietary structure. As for lead, because the dose–response analyses did not provide any indication of a threshold for the key effects of lead, JECFA concluded that a guideline of tolerable intake that would be considered to be health protective was not possible to establish [12].

Health risk assessment was the process that evaluated the potential health effects from doses of a contaminant delivered to humans in some manner. The health risks of trace element intake from the consumption of wine were assessed based on the THQ, which had been recognized as a useful parameter for the evaluation of risk associated with a contaminant. The THQ was the ratio of the estimated dose of elements from wines to a corresponding reference dose, and this approach offered an indication of the risk level due to contaminant exposure. Reference doses were obtained by conversion from USEPA data [35], and the reference doses of Cr, Mn, Ni, Zn, As, Mo, Cd and Se were 3.00, 140, 20.0, 300, 0.300, 5.00, 1.00 and 5.00 μg/kg per day, respectively. The estimated THQ values of the elements studied were shown in Table 4, including the THQ at the maximum intake (P95). The THQ values of each estimated element did not exceed 1 in all regions, suggesting that the exposed population would not experience significant health risks when ingesting these individual elements from daily consumption of 200 mL of wine. In addition, the higher THQ values indicated a higher probability of exposed risk, even if the THQ value was not greater than 1.

Table 3. Estimated daily intake (EDI, µg/kg bw/day) of each element via consumption of wine from eight Chinese regions.

Region [1]	Cr		Mn		Ni		Zn		As		Mo		Cd		Pb		Al		Se	
	Mean	P95	Mean	P95	Mean	P95	Mean	P95	Mean	P95	Mean	P95	Mean	P95	Mean	P95	Mean	P95	Mean	P95
NE	0.704	0.821	26.0	30.7	0.270	0.325	2.41	2.85	0.041	0.052	0.007	0.008	0.004	0.005	0.069	0.078	3.53	4.21	0.042	0.043
XJ	0.460	0.485	6.09	6.54	0.088	0.102	0.96	1.07	0.021	0.023	0.011	0.013	0.001	0.001	0.023	0.025	2.21	2.59	0.034	0.035
HM	0.437	0.462	6.23	6.65	0.092	0.102	1.26	1.37	0.020	0.022	0.006	0.007	0.001	0.001	0.038	0.041	2.74	3.06	0.031	0.032
HC	0.466	0.498	6.55	7.17	0.095	0.108	1.31	1.52	0.020	0.022	0.007	0.009	0.001	0.002	0.037	0.043	2.89	3.71	0.032	0.035
LP	0.443	0.483	7.43	8.10	0.099	0.116	1.78	2.16	0.013	0.016	0.006	0.009	0.002	0.002	0.059	0.069	3.52	4.44	0.028	0.030
YV	0.410	0.436	6.87	7.57	0.075	0.084	1.41	1.57	0.015	0.016	0.005	0.007	0.001	0.001	0.045	0.049	2.53	2.78	0.029	0.030
BG	0.503	0.528	20.8	24.1	0.242	0.283	1.99	2.21	0.022	0.029	0.005	0.005	0.004	0.004	0.084	0.094	3.30	3.61	0.030	0.032
SWH	0.393	0.424	7.61	8.32	0.078	0.088	1.49	1.72	0.014	0.017	0.005	0.006	0.002	0.003	0.045	0.053	1.84	2.30	0.028	0.029
PTDI			183 [2]		12.0		1000		2.14 [3]				0.833 [4]				143 [5]		6.67 [6]	

[1] NE: Northeast; XJ: Xinjiang; HM: Helan Mountain; HC: Hexi Corridor; LP: Loess Plateau; YV: Yanhuai Valley; BG: Bohai Gulf; SWH: Southwest Highland. [2] The PTDI (µg/kg bw/day) of Mn was calculated from a tolerable daily intake of 11.0 mg/day by a 60 kg adult. [3] The PTDI (µg/kg bw/day) of As was calculated from a provisional tolerable weekly intake (PTWI) of 15 µg/kg body weight. [4] The PTDI (µg/kg bw/day) of Cd was calculated from a provisional tolerable monthly intake (PTMI) of 25.0 µg/kg body weight based on a month of 30 days. [5] The PTDI (µg/kg bw/day) of Al was calculated from PTWI of 1000 µg/kg body weight. [6] The PTDI (µg/kg bw/day) of Se was calculated from an upper tolerable limit for selenium of 400 µg/day by a 60 kg adult.

Table 4. Target hazard quotient (THQ) of trace elements from wines of different regions.

Elements	NE		XJ		HM		HC		LP		YV		BG		SWH	
	Mean	P95	Mean	P95	Mean	P95	Mean	P95	Mean	P95	Mean	P95	Mean	P95	Mean	P95
Cr	0.235	0.274	0.153	0.162	0.146	0.154	0.155	0.166	0.148	0.161	0.137	0.145	0.168	0.176	0.131	0.141
Mn	0.186	0.220	0.044	0.047	0.044	0.048	0.047	0.051	0.053	0.058	0.049	0.054	0.148	0.172	0.054	0.059
Ni	0.014	0.016	0.004	0.005	0.005	0.005	0.005	0.005	0.005	0.006	0.004	0.004	0.012	0.014	0.004	0.004
Zn	0.008	0.009	0.003	0.004	0.004	0.005	0.004	0.005	0.006	0.007	0.005	0.005	0.007	0.007	0.005	0.006
As	0.137	0.173	0.070	0.077	0.067	0.073	0.067	0.073	0.043	0.053	0.050	0.053	0.073	0.097	0.047	0.057
Mo	0.001	0.002	0.002	0.003	0.001	0.001	0.001	0.002	0.001	0.002	0.001	0.001	0.001	0.001	0.001	0.001
Cd	0.004	0.005	0.001	0.001	0.001	0.001	0.001	0.002	0.002	0.002	0.001	0.001	0.004	0.004	0.002	0.003
Se	0.008	0.009	0.007	0.007	0.006	0.006	0.006	0.007	0.006	0.006	0.006	0.006	0.006	0.006	0.006	0.006
Total	0.593	0.708	0.284	0.306	0.274	0.293	0.286	0.311	0.264	0.295	0.253	0.269	0.419	0.477	0.250	0.277

Notes: NE: Northeast; XJ: Xinjiang; HM: Helan Mountain; HC: Hexi Corridor; LP: Loess Plateau; YV: Yanhuai Valley; BG: Bohai Gulf; SWH: Southwest Highland.

When all estimated elements were taken into account, the total THQ values performed to estimate the cumulative health risk effect were calculated by sum of THQ values of these elements, which was customarily called the HI. In the table, the HI values of different regions were also less than 1, and the highest HI values were 0.593 and 0.708 (P95), which were found in the Northeast region. Furthermore, the HI values in regions other than the Bohai Gulf region (0.419 and 0.477 for P95) were mostly similar to each other but less than those of the Bohai Gulf region. It showed that these contaminant elements did not cause health hazards to people through the daily consumption of these quantities of wine individually, but the risks that might present were higher in the wine of the Northeast region than in the wine of the other regions. For all the estimated wines, the THQ values of Cr accounted for the highest average proportions of the total THQ values, 48.5% and 47.0% (P95), and the average contribution proportions of Mn to the total THQ values which were 23.8% and 24.2% (P95) ranked second. On the other hand, the THQ also represented the contribution of wine to contaminants in the acceptable range for the daily diet in this study. For example, the average THQ of Cr in the measured wines was 0.159, which meant that the contribution of wine consumption to the tolerable daily intake of Cr was 15.9%. Risk assessment for a specific contaminant intake required comprehensive consideration of all intake pathways, and wine consumption was just one such path. Therefore, the proportion of wine consumption was more important for health risk assessment of wine in the daily diet of drinkers.

3. Materials and Methods

3.1. Study Area and Sampling

The study area consisted of eight major Chinese wine-producing regions, including the eastern foothills of Helan Mountain (HM), Xinjiang (XJ), the Hexi Corridor (HC), the ring around Bohai Gulf (BG), Southwest Highland (SWH), Yanhuai Valley (YV), the Northeast (NE) and the Loess Plateau (LP) (Figure 1).

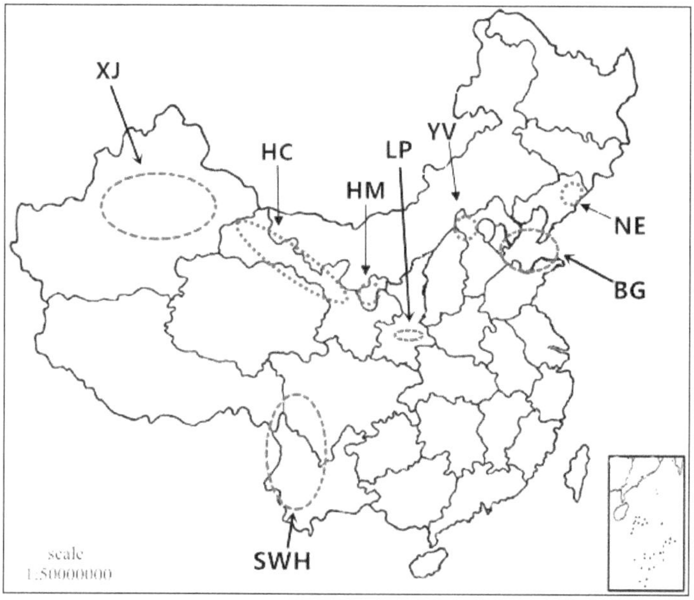

Figure 1. Geographical location of study areas of wine production in China (HM: Helan Mountain; XJ: Xinjiang; HC: Hexi Corridor; BG: Bohai Gulf; SWH: Southwest Highland; YV: Yanhuai Valley; NE: Northeast; LP: Loess Plateau).

Wine samples with vintages all between 2013 and 2014 were collected from a total of 71 local different chateaus or wineries in 2015, which chateaus or wineries had the Protected Geographical Indication (PGI) for wines. All wine samples were taken directly from wine tanks or oak barrels, and the varieties and detailed quantities of analyzed wines are shown in Table 5.

Table 5. The varieties and number of 315 wines in different regions.

Region	Varieties	Number of Wines
NE	Beibinghong, Vidal	30
XJ	Cabernet Sauvignon, Merlot	50
HM	Cabernet Sauvignon, Merlot	60
HC	Cabernet Sauvignon, Merlot	30
LP	Cabernet Sauvignon, Cabernet Franc	20
YV	Cabernet Sauvignon, Syrah	35
BG	Cabernet Gernischet, Marselan	40
SWH	Rose Honey, Crystal	50

Notes: NE: Northeast; XJ: Xinjiang; HM: Helan Mountain; HC: Hexi Corridor; LP: Loess Plateau; YV: Yanhuai Valley; BG: Bohai Gulf; SWH: Southwest Highland.

3.2. Chemicals and Reagents

Metal-oxide-semiconductor grade concentrated nitric acid (HNO_3) was obtained from the Beijing Institute of Chemical Reagents (Beijing, China). Deionized water (18.2 MΩ cm) from a Milli-Q system (Millipore, Milford, MA, USA) was used throughout the experiments. A standard solution of trace elements, including aluminum (Al), arsenic (As), cadmium (Cd), chromium (Cr), cobalt (Co), copper (Cu), lead (Pb), manganese (Mn), molybdenum (Mo), nickel (Ni), selenium (Se) and zinc (Zn), at a concentration of 100 mg/L was purchased from Seigniory Chemical Products Ltd. (SCP Science, Montreal, QC, Canada). A mixed tuning solution containing magnesium (Mg), indium (In), cerium (Ce), barium (Ba) and uranium (U) at a concentration of 1.0 μg/L was obtained from PerkinElmer Corporation (Waltham, MA USA). The solution with yttrium (Y) as internal standard was purchased from Seigniory Chemical Products Ltd. Certified rice reference material (GBW 10010) was purchased from the National Standard Substance Research Center (NSSRC, Beijing, China).

3.3. Preparation and Analytical Methods

An Anton Paar microwave oven (Anton Paar Multiwave 3000, Anton Paar GmbH, Graz, Austria) was used for all sample digestions. All vessels in the whole study were soaked in 30% HNO_3 for one night and then rinsed with deionized water more than three times. Five milliliters of wine sample were placed in a PTFE vessel. The vessel was placed on an electric hot plate (LabTech EHD 36, Beijing LabTech Corporation, Beijing, China) at 100 °C to evaporate the ethanol until the wine sample was concentrated to 2 mL volume. Then, 1 mL of concentrated HNO_3 was added, and the sample was subjected to microwave digestion. The conditions of the microwave-assisted digestion were as follows: (1) temperature: 190 °C; (2) pressure: 200 psi (1 psi = 6890 Pa); (3) climbing time: 20 min; and (4) retention time: 10 min. This procedure was completed in a closed system, and the samples did not contact the outside environment. The colorless transparent liquid in the vessel indicated that digestion was accomplished. When the process finished, the vessel was retained on an electric hot plate at 100 °C to evaporate redundant acid until the sample was concentrated to 2 mL volume. The vessel was cooled to room temperature and then moved into a volumetric flask after cooling. Finally, the digested solution was constant volume to 25 mL with 2% HNO_3 and waiting for determination. Reagent blanks and matrix spike duplicates were also treated in accordance with the abovementioned procedure.

The element concentrations were determined using ICP-MS (PerkinElmer ICP-MS Elan DRC-e, PerkinElmer Corporation) with a 40.68 MHz self-excited radio frequency generator, GemClean^{TM} cross nebulizer and Ryton^{TM} double channel atomizing chamber of highly inert polymer material. The main optimized instrumental parameters were as follows: (1) radio frequency power: 1100 W; (2) plasma gas flow rate: 15 L/min; (3) carrier gas flow rate: 0.94 L/min; (4) auxiliary gas flow rate: 1.2 L/min; (5) lens voltage: 5.5 V; and (6) sampling flow rate: 0.8 mL/min; (7) detection mode: standard mode. The correction equation recommended by the instrument software was used.

3.4. Quality Control

Quality assurance measures comprised analyzing matrix blank samples and the internal standard solution of each batch, and the certified reference material (GBW 10010) were inserted into the sample sequence every 10 samples to verify sensitivity and repeatability. The calibration curve of multi-element mixed standard solution was prepared from 100 mg/L standard solution of trace elements in 2% HNO_3, where the linear range of Al was 10.0–250 μg/L and the linear ranges of As, Cd, Cr, Co, Cu, Pb, Mn, Mo, Ni, Se, Zn were 0.250–50.0 μg/L. In addition, spike recovery tests were done on wine samples, and the spiked samples were prepared at three different concentration levels. The analytical precision and accuracy were accepted only when relative standard deviation (RSD) values were below 5% for the elements, according to the results of duplicate measurements of all samples and the certificated reference materials.

3.5. Statistical Analysis and Calculations

All analyses were performed in triplicate, the mean values of three duplicates were used as the result of each wine analysis. Duncan's multiple range tests were used to determine significant difference ($p < 0.05$) with SPSS 19.0 software for Windows (SPSS Inc., Chicago, IL, USA). The boxplot was created by SPSS 19.0.

The estimated daily intake (EDI, μg/kg bw/day) of trace elements from wine consumption depended on the concentration of the elements in the wine, daily consumption rate of wine and body weight of the consumers [36]:

$$\mathrm{EDI} = (\mathrm{C} \times \mathrm{R})/\mathrm{BW} \tag{1}$$

where C—the concentration of an individual element in the wine, μg/L; R—the daily wine consumption rate for adult wine drinkers, L/day; BW—the average body weight of assessed population, kg.

The health risk due to the consumption of wine was assessed based on the target hazard quotient (THQ), which was calculated from the ratio of EDI and an oral reference dose [37]:

$$\mathrm{THQ} = \mathrm{EDI}/\mathrm{RfD} \tag{2}$$

where RfD—the oral reference dose for each element, μg/kg bw/day.

If the THQ is less than 1, it means that the intake of this element from wine has no obvious adverse effects. If the THQ is equal to or higher than 1, there is a health risk to humans [38,39]. The total THQ values of the elements evaluated in wine is defined as hazard index (HI) used to evaluate the comprehensive health risks of wine.

4. Conclusions

The concentrations of Al, As, Cd, Cr, Co, Cu, Pb, Mn, Mo, Ni, Se and Zn in 315 wines sourced from the eight major Chinese wine-producing regions were determined, and the established method could accurately and effectively analyze the trace elements in wine. The results from this study suggested that the concentration levels of elements measured in wines decreased in the order of Mn > Al > Zn > Cu > Cr > Ni > Pb > Se > As > Co > Mo > Cd, and the concentrations of Mn and Cr were clearly higher than those in previous studies. The Northeast region had higher concentration level of these elements, and the Yanhuai Valley region had lower concentration level on the whole, which reflected the environmental condition and element distribution in these regions, with differences for different regions.

The estimated daily intake of elements with potential health risks from wine consumption, assuming a daily wine consumption of 200 mL for 60 kg drinkers, was far lower than the related tolerable daily intake for this element, including the possible highest intake. The health risk assessment indicated that the THQ of each evaluated element was all far below 1, which meant that the exposed population would not experience significant health risks from daily consumption of 200 mL of wines individually in ingesting these elements. The HI for wine consumption in each region was also less than 1, which again showed that the contribution of wine consumption did not pose a threat to tolerable daily intake of potential risk elements in the daily diet. However, the THQ values of Cr and Mn accounted for a larger proportion of the HI, which meant that they were major contaminants in the consumption of Chinese wine and needed more attention. On the other hand, the proportion of wine consumption in the daily diet of drinkers was more important for comprehensive health risk assessment of contaminants due to the presence of other intake pathways. Therefore, the cumulative impact of wine consumption on trace elements intake in the daily diet of drinkers should not be ignored.

Supplementary Materials: Figure S1: Concentration range and distribution of chromium (Cr) in wines from eight different regions, Figure S2: Concentration range and distribution of manganese (Mn) in wines from eight different regions, Figure S3: Concentration range and distribution of cobalt (Co) in wines from eight different regions, Figure S4: Concentration range and distribution of nickel (Ni) in wines from eight different regions, Figure S5: Concentration range and distribution of copper (Cu) in wines from eight different regions, Figure S6: Concentration range and distribution of copper zinc (Zn) in wines from eight different regions, Figure S7: Concentration range and distribution of arsenic (As) in wines from eight different regions, Figure S8: Concentration range and distribution of molybdenum (Mo) in wines from eight different regions, Figure S9: Concentration range and distribution of cadmium (Cd) in wines from eight different regions, Figure S10: Concentration range and distribution of lead (Pb) in wines from eight different regions, Figure S11: Concentration range and distribution of aluminum (Al) in wines from eight different regions, Figure S12: Concentration range and distribution of selenium (Se) in wines from eight different regions, Table S1: Determination results of trace elements in 315 wines (μg/L).

Author Contributions: Conceptualization, Z.-H.D. and Y.-L.F.; methodology, Z.-H.D. and A.Z.; software, Z.-W.Y.; validation, Z.-H.D., Y.-L.Z., and F.W.; investigation, J.M. and Y.-X.L.; resources, J.-J.Z.; data curation, Z.-H.D. and A.Z.; writing—original draft preparation, Z.-H.D.; writing—review and editing, A.Z.; supervision, Y.-L.F.; project administration, A.Z.; funding acquisition, J.-J.Z.. All authors have read and approved the final manuscript.

Funding: This research was funded by [the National Key Research and Development Program of China], grant number [2017YFC1601700] and The APC was funded by [the National Key Research and Development Program of China]. This work was supported by the National Key Research and Development Program of China (grant no. 2017YFC1601700) and General Administration of Quality Supervision, Inspection and Quarantine (AQSIQ) Science and Technology Plan (grant no. 2017IK253).

Conflicts of Interest: The authors declare no conflict of interest.

References

1. OIV. 2017 Global Economic Vitiviniculture Data. 2017. Available online: http://www.oiv.int/public/medias/5686/ptconj-octobre2017-en.pdf (accessed on 27 September 2018).
2. OIV. State of the Vitiviniculture World Market. 2018. Available online: http://www.oiv.int/public/medias/5958/oiv-state-of-the-vitiviniculture-world-market-april-2018.pdf (accessed on 27 September 2018).
3. Goldberg, I.J.; Mosca, L.; Piano, M.R.; Fisher, E.A. Wine and your heart: a science advisory for healthcare professionals from the Nutrition Committee, Council on Epidemiology and Prevention, and Council on Cardiovascular Nursing of the American Heart Association. *Circulation* **2001**, *103*, 472–475. [CrossRef] [PubMed]
4. Fulvio, U.; Alex, S. Wine polyphenols and optimal nutrition. *Ann. N.Y. Acad. Sci.* **2002**, *957*, 200–209.
5. Shrikhande, A.J. Wine by-products with health effects. *Food Res. Int.* **2000**, *33*, 469–474. [CrossRef]
6. Zuniga-Lopez, M.C.; Laurie, V.F.; Barriga-Gonzalez, G.; Folch-Cano, C.; Fuentes, J.; Agosin, E.; Olea-Azar, C. Chemical and biological properties of phenolics in wine: Analytical determinations and health benefits. *Curr. Org. Chem.* **2017**, *21*, 357–367. [CrossRef]
7. Martin, M.A.; Goya, L.; Ramos, S. Protective effects of tea, red wine and cocoa in diabetes. Evidences from human studies. *Food Chem. Toxicol.* **2017**, *109*, 302–314. [CrossRef] [PubMed]

8. Hughes, M.F.; Beck, B.D.; Chen, Y.; Lewis, A.S.; Thomas, D.J. Arsenic exposure and toxicology: a historical perspective. *Toxicol. Sci.* **2011**, *123*, 305–332. [CrossRef] [PubMed]
9. FAO/WHO. *Evaluation of Certain Food Additives and Contaminants, Thirty-third Report of the Joint FAO/WHO Expert Committee on Food Additives*; World Health Organization: Geneva, Switzerland, 1989.
10. Needleman, H.L. Leadpoisoning. *Ann. Rev. Med.* **2004**, *55*, 209–222. [CrossRef]
11. Fang, Y.; Sun, X.Y.; Yang, W.J.; Ma, N.; Xin, Z.H.; Fu, J.; Liu, X.C.; Liu, M.; Mariga, A.M.; Zhu, X.F.; Hu, Q.H. Concentrations and health risks of lead, cadmium, arsenic, and mercury in rice and edible mushrooms in China. *Food Chem.* **2014**, *147*, 147–151. [CrossRef]
12. FAO/WHO. *Evaluation of Certain Food Additives and Contaminants, Seventy-Third Report of the Joint FAO/WHO Expert Committee on Food Additives*; World Health Organization: Geneva, Switzerland, 2011.
13. Tuzen, M. Toxic and essential trace elemental contents in fish species from the Black Sea, Turkey. *Food Chem. Toxicol.* **2009**, *47*, 1785–1790. [CrossRef]
14. Forti, E.; Salovaara, S.; Cetin, Y.; Bulgheroni, A.; Tessadri, R.; Jennings, P.; Pfaller, W.; Prieto, P. In vitro evaluation of the toxicity induced by nickel soluble and particulate forms in human airway epithelial cells. *Toxicol. Vitro* **2011**, *25*, 454–461. [CrossRef]
15. WHO. Nickel in drinking-water. In *Background Document for Development of WHO Guidelines for Drinking-Water Quality*; World Health Organization: Geneva, Switzerland, 2005.
16. Koreňovská, M.; Suhaj, M. Identification of some Slovakian and European wines origin by the use of factor analysis of elemental data. *Eur. Food Res. Technol.* **2005**, *221*, 550–558. [CrossRef]
17. Meeker, J.D.; Rossano, M.G.; Protas, B.; Diamond, M.P.; Puscheck, E.; Daly, D.; Paneth, N.; Wirth, J.J. Cadmium, lead, and other metals in relation to semen quality: Human evidence for molybdenum as a male reproductive toxicant. *Environ. Health Persp.* **2008**, *116*, 1473–1479. [CrossRef] [PubMed]
18. FAO/WHO. *Evaluation of Certain Food Additives and Contaminants, Twenty-Sixth Report of the Joint FAO/WHO Expert Committee on Food Additives*; World Health Organization: Geneva, Switzerland, 1982.
19. FAO/WHO. *Evaluation of Certain Food Additives and Contaminants, Sixty-Seventh Report of the Joint FAO/WHO Expert Committee on Food Additives*; World Health Organization: Geneva, Switzerland, 2007.
20. WHO. Manganese in Drinking-Water. In *Background Document for Preparation of WHO Guidelines for Drinking-Water Quality*; World Health Organization: Geneva, Switzerland, 2011.
21. FAO/WHO. *Vitamin and Mineral Requirements in Human Nutrition*, 2nd ed.; Report of a Joint FAO/WHO Expert Consultation, Bangkok, Thailand, 21–30 September 1998; World Health Organization: Geneva, Switzerland, 2004; Available online: http://whqlibdoc.who.int/publications/2004/9241546123.pdf (accessed on 27 September 2018).
22. Meharg, A.A.; Rahman, M.M. Arsenic contamination of Bangladesh paddy field soils: Implications for rice contribution to arsenic consumption. *Environ. Sci. Technol.* **2003**, *37*, 229–234. [CrossRef] [PubMed]
23. Khan, M.U.; Malik, R.N.; Muhammad, S. Human health risk from heavy metal via food crops consumption with wastewater irrigation practices in Pakistan. *Chemosphere* **2013**, *93*, 2230–2238. [CrossRef] [PubMed]
24. Fiket, Ž.; Mikac, N.; Kniewald, G. Arsenic and other trace elements in wines of eastern Croatia. *Food Chem.* **2011**, *126*, 941–947. [CrossRef]
25. Sperkova, J.; Suchanek, M. Multivariate classification of wines from different bohemian regions (Czech Republic). *Food Chem.* **2005**, *93*, 659–663. [CrossRef]
26. Marengo, E.; Aceto, M. Statistical investigation of the differences in the distribution of metals in Nebbiolo-based wines. *Food Chem.* **2003**, *81*, 621–630. [CrossRef]
27. Gremaud, G.; Quaile, S.; Piantini, U.; Pfammatter, E.; Corvi, C. Characterization of swiss vineyards using isotopic data in combination with trace elements and classical parameters. *Eur. Food Res. Technol.* **2004**, *219*, 97–104. [CrossRef]
28. Geana, I.; Iordache, A.; Ionete, R.; Marinescu, A.; Ranca, A.; Culea, M. Geographical origin identification of Romanian wines by ICP-MS elemental analysis. *Food Chem.* **2013**, *138*, 1125–1134. [CrossRef]
29. Alkış, İ.M.; Öz, S.; Atakol, A.; Yılmaz, N.; Anlı, R.E.; Atakol, O. Investigation of heavy metal concentrations in some Turkish wines. *J. Food Compos. Anal.* **2014**, *33*, 105–110. [CrossRef]
30. OIV. Maximum Acceptable Limits of Various Substances Contained in Wine. 2011. Available online: http://www.oiv.int/public/medias/2601/oiv-ma-c1-01.pdf (accessed on 27 September 2018).
31. State Food and Drug Administration. *Limit of Pollutants in Food*; National Health and Family Planning Commission of the People's Republic of China: Beijing, China, 2017.

32. Towle, K.M.; Garnick, L.C.; Monnot, A.D. A human health risk assessment of lead (pb) ingestion among adult wine consumers. *Int. J. Food Contamination* **2017**, 4. [CrossRef]
33. WHO. Chromium in drinking-water. In *Background Document for Preparation of WHO Guidelines for Drinking-Water Quality*; World Health Organization: Geneva, Switzerland, 2003.
34. WHO. Molybdenum in drinking-water. In *Background Document for Preparation of WHO Guidelines for Drinking-Water Quality*; World Health Organization: Geneva, Switzerland, 2011.
35. USEPA. Integrated Risk Information System Assessments. Available online: https://cfpub.epa.gov/ncea/iris2/atoz.cfm (accessed on 27 September 2018).
36. Qian, Y.Z.; Chen, C.; Zhang, Q.; Li, Y.; Chen, Z.J.; Li, M. Concentrations of cadmium, lead, mercury and arsenic in Chinese market milled rice and associated population health risk. *Food Control* **2010**, *21*, 1757–1763. [CrossRef]
37. USEPA. Risk Assessment Guidance for Superfund, Volume I. In *Human Health Evaluation Manual, Part A*; United States Environmental Protection Agency Publication: Washington, DC, USA, 1989.
38. Wang, X.L.; Sato, T.; Xing, B.S.; Tao, S. Health risks of heavy metals to the general public in Tianjin, China via consumption of vegetables and fish. *Sci. Total Environ.* **2005**, *350*, 28–37. [CrossRef] [PubMed]
39. Zheng, N.; Wang, Q.C.; Zhang, X.W.; Zheng, D.M.; Zhang, Z.S.; Zhang, S.Q. Population health risk due to dietary intake of heavy metals in the industrial area of Huludao city, China. *Sci. Total Environ.* **2007**, *387*, 96–104. [CrossRef] [PubMed]

Sample Availability: Samples of the compounds are not available from the authors.

molecules

MDPI

Article

Use of Multiflora Bee Pollen as a Flor Velum Yeast Growth Activator in Biological Aging Wines

Pau Sancho-Galán, Antonio Amores-Arrocha *, Ana Jiménez-Cantizano and Víctor Palacios

Department of Chemical Engineering and Food Technology, Faculty of Sciences, University of Cadiz, Agrifood Campus of International Excellence (ceiA3), IVAGRO, P.O. Box 40, 11510 Puerto Real, Cadiz, Spain; pau.sancho@uca.es (P.S.-G.); ana.jimenezcantizano@uca.es (A.J.-C.); victor.palacios@uca.es (V.P.)

* Correspondence: antonio.amores@uca.es; Tel.: +34-9-5601-6398

Academic Editors: Rosa Maria de Sá Perestrelo and José Sousa Câmara

Received: 11 April 2019; Accepted: 4 May 2019; Published: 7 May 2019

Abstract: Flor velum yeast growth activators during biological aging are currently unknown. In this sense, this research focuses on the use of bee pollen as a flor velum activator. Bee pollen influence on viable yeast development, surface hydrophobicity, and yeast assimilable nitrogen has already been studied. Additionally, bee pollen effects on the main compounds related to flor yeast metabolism and wine sensory characteristics have been evaluated. "Fino" (Sherry) wine was supplemented with bee pollen using six different doses ranging from 0.1 to 20 g/L. Its addition in a dose equal or greater than 0.25 g/L can be an effective flor velum activator, increasing yeast populations and its buoyancy due to its content of yeast assimilable nitrogen and fatty acids. Except for the 20 g/L dose, pollen did not induce any significant effect on flor velum metabolism, physicochemical parameters, organic acids, major volatile compounds, or glycerol. Sensory analysis showed that low bee pollen doses increase wine's biological aging attributes, obtaining the highest score from the tasting panel. Multiflora bee pollen could be a natural oenological tool to enhance flor velum development and wine sensory qualities. This study confirms association between the bee pollen dose applied and the flor velum growth rate. The addition of bee pollen could help winemakers to accelerate or reimplant flor velum in biologically aged wines.

Keywords: Bee pollen; biological aging; activator; sherry wine

1. Introduction

Biological aging is a microbiological process where a natural biofilm of a *Saccharomyces* yeast strain called flor velum develops on wine surface, resulting in Fino or Manzanilla wines [1,2]. Also, in some wine producing areas of the world such as Sardinia (Italy), Jura (France), Montilla-Moriles (Spain), and Tokay (Hungry), special wines are produced by biological aging processes [3]. Sherry flor velum are a film forming culture 95% composed of different *Saccharomyces cerevisiae* strains, growing under hard conditions such as low oxygen concentrations, high ethanol concentrations (from 15% to 16% *v*/*v*), low pH levels, and low fermentable sugar concentrations [4,5]. The flor velum develops an oxidative metabolism, where ethanol consumption as a primary carbon source has been widely studied, as well as glycerol, acetic, and lactic acid [6]. In this way, acetaldehyde, 2,3-butanediol, fusel alcohols, diacetyl, and acetoin among others are formed as reaction products [2,4,7,8]. For cell development and protein synthesis [9–11], flor velum yeasts use nutrients and growth factors such as nitrogen compounds, fatty acids, vitamins, etc., taken mainly from wine [12–15]. Scarce studies have been conducted on the relationship of the wine nitrogen content and composition, and the development of flor velum yeasts. Some authors highlighted the importance of amino acid consumption and production for the cellular redox potential balance [16]. In addition, Berlanga et al. (2006) [10] studied different processes such as the amino acid conversion into other more reduced, major alcohols, acids, and other compounds during biological aging. Flor velum yeasts can synthesize or release new amino acids during the biological aging process [16], and L-proline is the main source of nitrogen for flor velum yeasts during biological aging [6]. Nowadays, there is only one technique to activate the flor velum growth, and is based on the aeration and introduction of new wine in oak casks in order to introduce oxygen and nutrients [17]. Furthermore, in the current oenological market, there is a wide variety of amino acids and ammonium based products designed to supply grape musts with nutritional deficiencies [11,18], but there are not such products for the biological aging process.

In this regard, it should be noted that multiflora bee pollen could be a natural product that could fill the biological aging activation gap. Bee pollen is a natural product that comes from beehives and is rich in carbohydrates, lipids, amino acids, proteins, minerals, fatty acids, polyphenols, sterols, and phospholipids among other compounds of interest [19–21]. Previous research works showed how the application of high multiflora bee pollen doses produced a significant yeast assimilable nitrogen (YAN) increase in mead and white grape musts [18,22], decreasing the yeast lag phase during the alcoholic fermentation. This may be due to the fact that YAN is essential for the optimal growth and development of yeasts and it includes amoniacal nitrogen, amino acids, small peptides, and nitrogen that can be easily assimilated by yeasts [23].

For these reasons, the aim of this study is to evaluate the use of multiflora bee pollen as an activator of the biological aging process under flor velum and its influence on viable yeast population development, surface hydrophobicity or ability to float, and wine YAN content. In addition, the main parameters and compounds related to flor velum yeast metabolism and the typical sensory attributes of wines supplemented with pollen have been assessed.

2. Results

2.1. *Influence of Bee Pollen on Flor Velum Yeast Development and Yeast Assimilable Nitrogen (YAN)*

Figure 1 shows flor velum viable yeast evolution during the biological aging in multiflora bee pollen supplemented wines. Except for 0.1 g/L dose, bee pollen addition increased significantly total viable yeast cells in respect to the control, and a linear correlation between the bee pollen dose applied and the maximum viable yeast cells was observed ($R^2 = 0.94$). As it can be seen, from the 5 g/L bee pollen dose on, flor velum yeast latency time was significantly reduced (3 days) compared to the control (5 days) (Figure 1). Both the control and 0.1 g/L dose samples presented a similar behavior, with a constant linear growth ($R^2 = 0.984$ and $R^2 = 0.964$ for the control and 0.1 g/L dose, respectively). At the 0.25 g/L dose the evolution was also linear, although a greater slope was observed in respect to the control and 0.1 g/L dose. Nevertheless, from the 5 g/L bee pollen dose on, an exponential yeast growth was observed until the 15th–18th days. After those days, the yeast growing rate slowed gradually until the end of the experiment. From the 18th day onwards, all doses except 0.1 g/L showed significantly higher population values in respect to the control (ANOVA $p < 0.05$), hence, the concentration of viable flor velum yeasts was 100–200% higher than the control for doses ranging between 0.25–20 g/L respectively. In this way, multiflora bee pollen was shown as an effective activator for development of flor velum yeasts from concentrations equal or greater than 0.25 g/L.

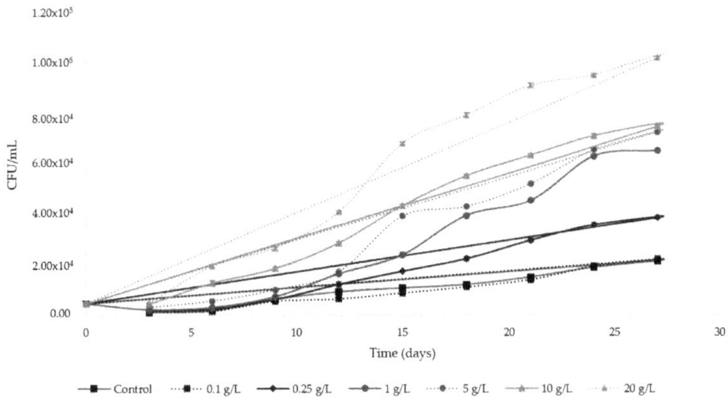

Figure 1. Evolution of the viable biomass of flor velum yeasts during the biological aging process of Palomino Fino wine with bee pollen doses. The results are the mean ± SD of three repetitions.

Figure 2 shows the evolution of YAN content in wines during biological aging. As it can be seen, the effect of bee pollen on the initial wine YAN concentration was remarkable, increasing linearly with the bee pollen dose applied ($R^2 = 0.98$). Once the biological aging process and yeast growth phase started, a significant YAN decrease was observed in all wines including the control, reaching the minimum YAN concentration at the 15th day (Figure 2). YAN content fall was more remarkable for the control and low-intermediate doses (0.1–1 g/L) with an average net reduction of 70%, than for medium or high doses (5–20 g/L), where the maximum reduction of YAN was less than 50%. This effect leads to the YAN consumption per colony forming unit (CFU) at the 15th–18th days decreasing significantly with the addition of bee pollen compared to control, with a maximum net reduction up to 60% for the 20 g/L dose (Figure 3). Once YAN values reached a minimum for each dose, its content started to increase, coinciding with the flor velum yeast stationary phase in most cases (Figure 1). At the end of the biological aging trials, all wines had a 37% lower YAN content than at the beginning of the experiment, and only the 10 and 20 g/L bee pollen doses showed significant differences (ANOVA $p < 0.05$) in respect to the control and the rest of the doses.

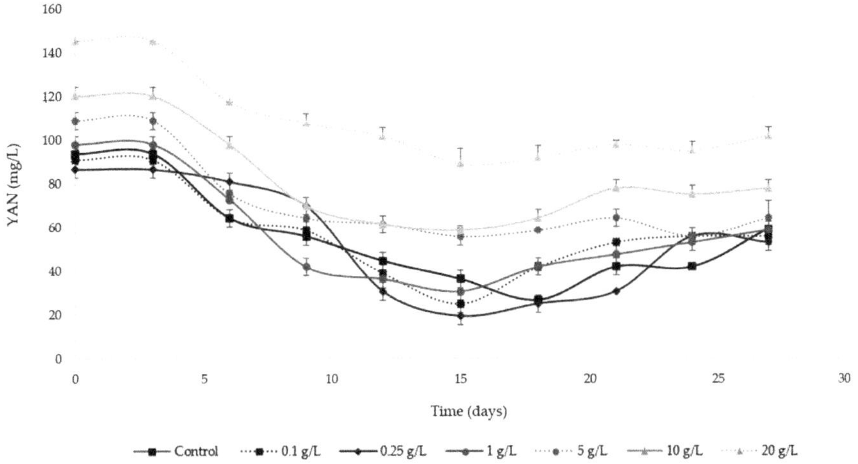

Figure 2. Evolution of the yeast assimilable nitrogen (YAN) by flor velum yeasts during the biological aging process of Palomino Fino wine with bee pollen doses. The results are the mean ± SD of three repetitions.

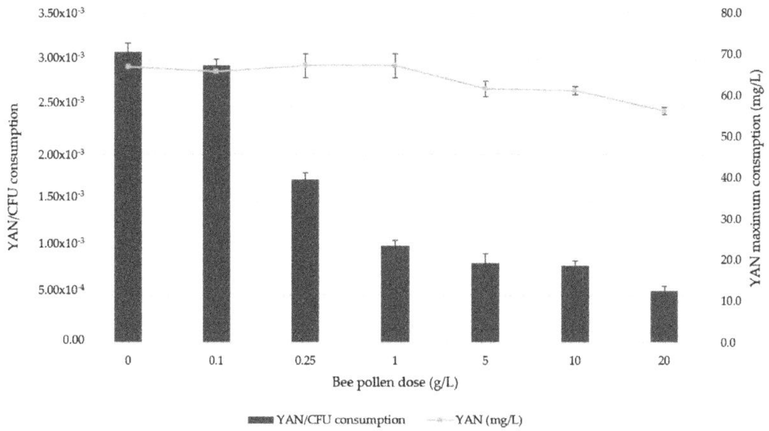

Figure 3. Maximum consumption of yeast assimilable nitrogen (YAN) per colony forming unit (CFU) during the exponential growth phase of Palomino Fino wine with different bee pollen doses and control.

2.2. *Influence of Bee Pollen on Flor Velum Hydrophobicity during Biological Aging*

Figure 4 shows flor velum yeast surface hydrophobicity evolution during biological aging. The evolution of surface hydrophobicity was related to the flor velum population development, and for this reason, cell surface hydrophobicity values could not be obtained until the 6th day. As expected, the hydrophobicity increase was exponential during the first 15 days for the pollen doses between 1 and 20 g/L, corresponding with the yeast exponential growth phase (Figure 1). Hydrophobicity levels reached in those doses were significantly higher (20–30%) compared to the control (ANOVA $p < 0.05$). Control and low bee pollen doses showed a linear growth trend, and the maximum values reached were not significantly different from wines with intermediate or high bee pollen doses (ANOVA $p < 0.05$). In addition, in the last days of biological aging, a decrease in the surface hydrophobicity values was observed, coinciding with the yeast stationary growth phase (Figure 1). Therefore, bee pollen addition (≥1 g/L) produces a significant increase in flor velum surface hydrophobicity during the exponential growth phase, giving stability and buoyancy to the velum.

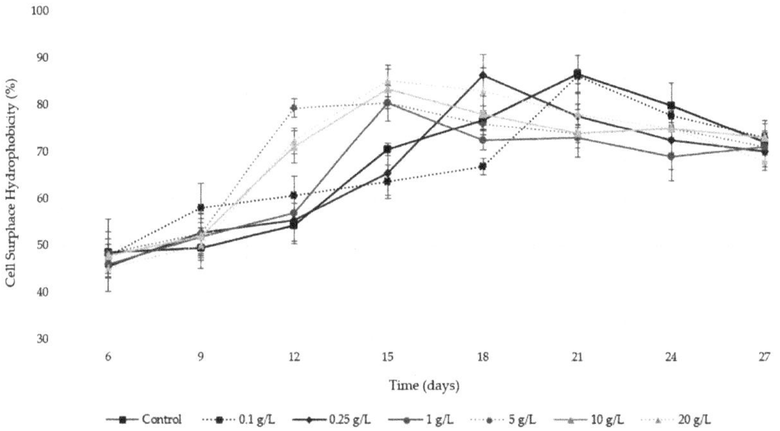

Figure 4. Cell surface hydrophobicity levels during the biological aging of Palomino Fino wine with bee pollen doses. The results are the mean ± SD of three repetitions.

2.3. Influence of Bee Pollen on Flor Velum Yeast Metabolism

Table 1 shows the influence of bee pollen addition on some compounds and parameters involved in flor velum metabolism. Regarding pH and total acidity no significant differences were found between the different bee pollen doses applied. Only a slight increase in pH for the 20 g/L bee pollen dose was observed. By comparing the ethanol values observed in Table 1 with those obtained by the potassium sorbate control, where no biological aging existed (13.38% ± 0.18), it can be determined that the flor velum yeast consumption only represents approximately 3.7% of the total decrease. Acetic acid represents another metabolite that flor velum consumes. As it can be seen in Table 1, volatile acidity presents an oscillatory behavior. Low and intermediate doses of bee pollen showed a significant decrease in volatile acidity values compared to the initial sample while for control and high doses (10–20 g/L) there was essentially no variation. Regarding organic acids composition, citric acid present in final wines ranged between 0.015 and 0.019 g/L (control and 20 g/L bee pollen, respectively). These values present a positive correlation between the dose of bee pollen applied and the concentration of this metabolite, with significant differences from the 5 g/L dose on (ANOVA, $p < 0.05$). In respect to tartaric acid values, these oscillated between 1.67 g/L (20 g/L bee pollen) and 2.02 g/L (control), with no significant differences between samples and control (Table 1). On one hand, malic acid did not show significant differences between the bee pollen samples and control. On the other hand, lactic acid content decreased significantly for all samples compared to the initial sample (ANOVA, $p < 0.05$). Finally, succinic acid was found in wines between 0.285 and 0.194 mg/L (control and the 20 g/L bee pollen, respectively). Although a 33% lower content than in the initial sample was observed, no significant differences were found between the different doses of bee pollen and the control. This last acid shows, unlike the previous ones, a negative correlation since it occurs at a lower concentration with a greater addition of bee pollen to wine.

Regarding the variation in the major volatile compounds content, a significant increase in acetaldehyde content was observed for all samples in respect to the initial simple. As it can be seen, acetaldehyde levels were significantly higher (between 15–53%) in wines supplemented with bee pollen in respect to the control, except for the highest dose (20 g/L) where the concentration reached was lower (Table 1). In relation to ethyl acetate concentration in wines, all the samples presented values between 52.71 mg/L (control) and 77.29 mg/L (20 g/L bee pollen); a significant relationship with acetic acid content was found [24]. In relation to methanol content, an increase of its concentration in respect to the control was observed showing significant differences from the 1 g/L dose on.

In respect to major alcohols (isobutanol, 1-propanol, and isoamyl alcohol) the same behavior was observed for the three alcohols studied up to the 5 g/L bee pollen dose sample. Low and intermediate bee pollen doses increased the major alcohol concentration in wines up to 52% compared to the initial sample. However, for the 20 g/L bee pollen dose no changes were observed in respect to the initial sample. Finally, glycerol content decreased significantly for all the samples in respect to the initial one. In addition, only the 20 g/L dose showed a significant decrease in respect to the rest of the wines after the biological aging.

Table 1. Effect of bee pollen on the physicochemical parameters and the total organic acids and major volatile compounds content.

Parameter	Before Biological Aging	After Biological Aging						
	Initial Sample	Control	0.1 g/L	0.25 g/L	1 g/L	5 g/L	10 g/L	20 g/L
pH	3.110 ± 0.014 ^{a}	2.870 ± 0.02 ^{a}	2.950 ± 0.01 ^{a}	2.970 ± 0.03 ^{a}	3.000 ± 0.15 ^{a}	3.000 ± 0.08 ^{a}	3.060 ± 0.01 ^{a}	3.160 ± 0.05 ^{a}
% Alcohol *v/v*	15.015 ± 0.064 ^{a}	12.555 ± 0.06 ^{b}	13.010 ± 0.06 ^{b}	13.010 ± 0.24 ^{b}	12.760 ± 0.015 ^{b}	12.790 ± 0.08 ^{b}	12.875 ± 0.12 ^{b}	12.975 ± 0.177 ^{b}
Total acidity (g/L)	5.870 ± 0.070 ^{a}	6.030 ± 0.040 ^{a}	5.990 ± 0.060 ^{a}	5.840 ± 0.060 ^{a}	5.840 ± 0.060 ^{a}	5.720 ± 0.080 ^{a}	5.920 ± 0.010 ^{a}	6.020 ± 0.050 ^{a}
Volatile acidity (g/L)	0.270 ± 0.050 ^{a}	0.300 ± 0.010 ^{a,d}	0.220 ± 0.010 ^{b}	0.200 ± 0.010 ^{b}	0.160 ± 0.010 ^{c}	0.270 ± 0.010 ^{a}	0.320 ± 0.010 ^{d}	0.370 ± 0.010 ^{e}
Citric acid (mg/L)		0.015 ± 0.001 ^{a}	0.015 ± 0.001 ^{a}	0.016 ± 0.001 ^{a}	0.016 ± 0.001 ^{a}	0.019 ± 0.001 ^{a,b}	0.024 ± 0.001 ^{b,c}	0.029 ± 0.001 ^{c}
Tartaric acid (g/L)	2.166 ± 0.001 ^{a}	2.018 ± 0.016 ^{a}	2.070 ± 0.002 ^{a}	2.025 ± 0.002 ^{a}	2.089 ± 0.016 ^{a}	1.977 ± 0.005 ^{a}	1.976 ± 0.007 ^{a}	1.677 ± 0.017 ^{a}
Malic acid (mg/L)	0.226 ± 0.001 ^{a}	0.240 ± 0.001 ^{a}	0.244 ± 0.001 ^{a}	0.240 ± 0.001 ^{a}	0.251 ± 0.003 ^{a}	0.253 ± 0.001 ^{a}	0.260 ± 0.002 ^{a}	0.273 ± 0.003 ^{a}
Succinic acid (mg/L)	0.450 ± 0.003 ^{a}	0.285 ± 0.001 ^{b}	0.272 ± 0.001 ^{b}	0.276 ± 0.002 ^{b}	0.263 ± 0.002 ^{b}	0.208 ± 0.002 ^{b}	0.199 ± 0.003 ^{b}	0.194 ± 0.001 ^{b}
Lactic acid (mg/L)	0.109 ± 0.005 ^{a}	0.013 ± 0.007 ^{b}	0.003 ± 0.001 ^{c}	0.010 ± 0.001 ^{b,c}	0.018 ± 0.001 ^{b,d}	0.021 ± 0.001 ^{d}	0.036 ± 0.001 ^{e}	0.024 ± 0.002 ^{d}
Acetaldehyde (mg/L)	77.648 ± 9.900 ^{a}	135.344 ± 2.806 ^{b}	184.024 ± 3.689 ^{c}	164.583 ± 8.479 ^{d}	207.665 ± 8.339 ^{e}	182.735 ± 6.958 ^{c}	155.533 ± 5.505 ^{d}	92.824 ± 2.135 ^{a}
Ethyl acetate (mg/L)	38.275 ± 2.167 ^{a}	52.705 ± 4.276 ^{b}	47.761 ± 1.450 ^{b}	50.462 ± 4.731 ^{b}	53.870 ± 6.109 ^{b}	53.304 ± 7.496 ^{b}	66.499 ± 4.032 ^{c}	77.285 ± 6.590 ^{d}
Methanol (mg/L)	35.039 ± 1.247 ^{a}	33.286 ± 2.702 ^{a}	43.978 ± 2.844 ^{b}	46.786 ± 2.468 ^{b}	53.726 ± 1.960 ^{c}	58.153 ± 1.249 ^{c}	57.755 ± 2.789 ^{c}	58.286 ± 1.099 ^{c}
1-Propanol (mg/L)	19.718 ± 1.287 ^{a}	29.463 ± 2.484 ^{b}	27.464 ± 4.640 ^{b}	34.342 ± 2.499 ^{c}	42.104 ± 2.475 ^{d}	42.596 ± 3.241 ^{d}	30.325 ± 2.221 ^{b}	20.662 ± 1.767 ^{a}
Isobutanol (mg/L)	32.222 ± 1.329 ^{a,d}	34.755 ± 1.734 ^{a,e}	43.989 ± 0.145 ^{b}	39.421 ± 1.496 ^{b,e}	58.307 ± 1.334 ^{c}	61.690 ± 1.775 ^{c}	29.241 ± 1.429 ^{d}	29.247 ± 2.484 ^{b}
Isoamyl alcohol (mg/L)	207.100 ± 7.838 ^{a}	215.534 ± 5.815 ^{a}	246.114 ± 1.160 ^{a,b}	256.654 ± 6.791 ^{a,b}	263.513 ± 4.338 ^{a,b}	290.323 ± 0.845 ^{b}	235.421 ± 3.881 ^{a,b}	201.665 ± 5.627 ^{a}
Glycerol (mg/L)	1624.240 ± 0.03 ^{a}	97.830 ± 3.34 ^{b}	99.910 ± 9.34 ^{b}	97.850 ± 6.67 ^{b}	100.410 ± 6.00 ^{b}	92.460 ± 7.34 ^{b}	83.300 ± 6.27 ^{b}	64.000 ± 6.00 ^{c}

Different uppercase letters mean statistically significant differences between samples at $p < 0.05$ obtained by two-way ANOVA and Bonferroni's multiple range (BSD) test. Results are the means ± SD of three repetitions.

2.4. *Descriptive Sensory Analysis*

To obtain a description of some of the wine organoleptic parameters, a sensory analysis of wines supplemented with pollen and the control was conducted. Spider web diagrams show the mean values for the attributes analyzed with the level of significance of control wine and samples with pollen. Figure 5 shows the results of visual and olfactory (a) and taste phase (b) respectively, as well as the global assessment of all wines tasted (b).

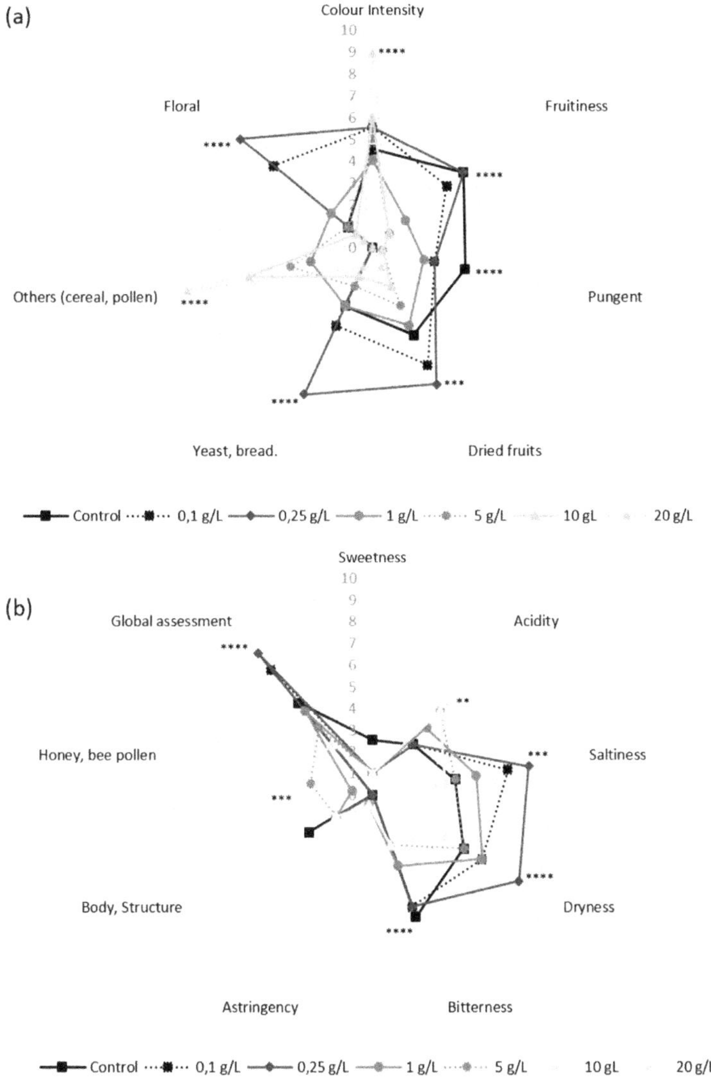

Figure 5. Bee pollen effect on the visual and olfactory (**a**) and taste (**b**) evaluation of Palomino Fino wine after biological aging. Stars indicate level of significance for two-way ANOVA according to Bonferroni's multiple range test (BSD) (** $p < 0.01$, *** $p < 0.001$ and **** $p < 0.0001$).

Significant differences were found for almost all the attributes studied. Color intensity showed significant differences with all the samples and the control for the 20 g/L bee pollen dose (Figure 5a).

Regarding floral aroma, only 0.1 and 0.25 g/L doses showed significant differences from the rest of the samples and the control. In respect to the pollen/cereal attributes, a positive correlation was found between the dose of pollen applied and the intensity of these attributes, and only from 1 g/L doses on, significant differences were found in respect to the control and low doses (0.1–0.25 g/L) (Figure 5a). Biological aging typical aromas such as yeast or dried fruit showed higher values for low bee pollen doses (0.1 and 0.25 g/L), and only the 0.25 g/L dose showed significant differences in respect to the control and the rest of the doses. In turn, attributes related to biologically aged wines, such as pungency, decreased with pollen contribution, showing its maximum intensity in control wine (Figure 5a). The opposite happens with the pollen/honey flavor perception (Figure 5b), which are significantly increased from 5 g/L doses on and affect wine sensory qualities. No significant differences were found for the sweetness, astringency, and body/structure attributes (Figure 5b). According to the global assessment results, low doses of bee pollen (0.1–0.25 g/L) have been found to be optimal for the sensory profile improvement, showing significant differences in respect to the control and the rest of the bee pollen doses (ANOVA, $p < 0.0001$).

3. Discussion

According to the results, pollen addition in doses equal or greater than 0.25 g/L produces a significant increase in flor velum yeast populations during biological aging (Figure 1). These results agree with those obtained by Amores-Arrocha et al. (2018) [18] for the alcoholic fermentation of white grape musts, where a linear correlation was also observed between the bee pollen doses applied and the maximum viable yeast cells observed ($R^2 = 0.86$). Results also agree with those obtained by Roldán et al. (2011) [22] for mead fermentation, where it was also concluded that bee pollen is a correct activator of the alcoholic fermentation process. The bee pollen activating effect on flor velum yeasts during biological aging may also be because it provides easily assimilable substances and growth factors such as amino acids, vitamins, and micronutrients [20]. Different authors indicate that bee pollen YAN is mainly due to its amino acids, representing 14–16% of YAN total content, especially proline, glutamic acid, aspartic acid, lysine, and leucine [20]. According to Roldán et al. (2011) [22], proline is the most important free amino acid in bee pollen composition and it is found in a 15.5 mg/g concentration. Dos Santos et al. (2000) [25] showed that during biological aging there is a reduction in the L-proline content since it is one of the main nitrogen sources consumed by flor velum yeasts. According to Berlanga et al. (2006) [10], glutamic acid and leucine can activate flor yeast development. All these amino acids may have a positive influence on flor velum development acting as a direct source for the synthesis of cellular proteins. Our results confirm that bee pollen addition implies a significant increase in YAN in wines at the beginning of the trial, and this fact may be one of the essential factors of flor velum yeast activation. However YAN consumption is not proportional to the maximum population reached during biological aging (Figure 3). It was expected that at high bee pollen doses, where a greater flor velum yeast growth was granted, YAN decrease should be greater. This fact could indicate that bee pollen could solubilize YAN throughout the process, especially in the initial growth phase, when the alcohol content is higher (15% approximately). Some authors have verified that YAN extraction in bee pollen increases with the alcoholic degree [18]. This effect was also observed during grape must alcoholic fermentation, but in this case, the maximum reduction reached with bee pollen addition was lower (20%) in respect to the control wine [18] and compared to the 60% achieved in our trials. In this way, this fact could indicate that bee pollen addition to wine could reduce the YAN requirements during biological aging since pollen supplies other growth-activating substances, such as fatty acids [26,27]. However, if is taken into account that pollen is a YAN reserve source, an increase in YAN content would be expected during the stationary phase for all the samples. This did not happen possibly because YAN consumption was still very high, due to an increase in flor velum populations and also because the YAN extraction effect is not favored as a result of a decrease in the alcoholic strength during this phase. Additionally, the YAN increase observed at the end of the biological aging could be favored by the yeast autolysis phenomenon in which an important amount of amino acids and other nitrogenized substances are released to the wine [28]. At the end of the biological aging, YAN content could be beneficial, for example, in case the flor velum has to be implanted again in a wine cask, in cases of cell death due to high temperatures, or to promote the flor velum development in very old Sherry wines, which have difficulties in keeping flor velum alive.

Regarding to the increase observed in hydrophobicity levels (Figure 4), this fact may be because multiflora bee pollen can provide some fatty acids responsible for the yeast cell wall hydrophobicity [29]. Aguilera et al. (1997) [30] found that oleic or palmitic acid contribution to the flor velum allows a greater cellular buoyancy. In this sense, some authors have verified that bee pollen has a high content of oleic and palmitic acids that represent approximately the 40% of the fatty acids total concentration [21,31]. Therefore, bee pollen is an input if substrates that allow the synthesis of compounds increasing cell surface hydrophobicity and the ability to float in a filmogenic phase over the wine.

In respect to the flor velum yeast metabolism (Table 1), the slight increase in pH for the 20 g/L bee pollen dose could be due to cation exchange, mainly potassium [21]. Ethanol (alcoholic strength) is the main metabolite consumed by flor velum yeasts [32], and it was expected to decrease at a greater rate, especially at high bee pollen doses (10–20 g/L) where the yeast populations were higher. However, this did not come into existence, even though the wine alcoholic strength for 10 and 20 g/L doses were higher than in the control. It is necessary to bear in mind that in our trials the ethanol decrease was due mainly to the evaporation phenomena that depend on many factors. Perhaps, velum with larger populations can wield a protective effect against the ethanol evaporation. Regarding volatile acidity, the slight variation observed for the samples may be a result of the limited growth of lactic acid bacteria present naturally in industrial velums [33,34] and/or production of acetic acid from acetaldehyde as a yeast metabolic regulator [35].

In relation to the organic acids composition, the presence of citric acid in such a low concentration is because it is a citric acid cycle substrate that is subsequently metabolized into ketoglutaric acid [7]. The slight tartaric acid concentration decrease may be associated with the pollen dose due to the potassium contribution by the pollen to the wine, which could cause the precipitation of this acid in the form of potassium salts [36]. Regarding the malic acid content, which is a fundamental substrate of lactic acid bacteria, no significant differences were observed between the different samples, and therefore it is deduced that the wines have not undergone any malolactic fermentation processes [37].

Referring to the major volatile compounds concentration, acetaldehyde is recognized as one of the main products in the biological aging process [38] and it is produced by the oxidative metabolism of ethanol [15]. However, its formation is influenced by the wine conditions [35] and especially by the major yeast strain present in the velum [39]. It has been found that the yeast strain employed in this research work is less tolerant than other *S. cerevisiae* strains to acetaldehyde, so when its concentration is high, generally above 200 mg/L, flor velum yeasts metabolize it through pathways like the Krebs or glyoxilate cycle. Thus, the acetaldehyde values presented by wines supplemented with pollen at the end of the biological aging are normal considering the *Saccharomyces cerevisiae* strain used in this study [38]. This yeast strain can also form acetic acid in order to reduce the medium toxicity. This fact may justify that the acetaldehyde levels in this study present oscillations depending on whether the yeast strain is in a predominant consumption or production phase. Regarding ethyl acetate concentration, bee pollen addition to wines is an input of fatty acids and amino acids [19]. In this way, bee pollen could be able to favor the esterification process to form ethyl acetate [40] through a major activation of acetyl transferase and esterase activities [41]. In respect to the methanol content, Amores-Arrocha et al. (2018) [42] obtained similar results. The increase observed could be associated with the high pectin content of bee pollen [22]. Some authors have confirmed that pectins can be solubilized and metabolized by yeasts during alcoholic fermentation, outputting methanol through an enzymatic hydrolysis [21]. In this way, is probably that this enzymatic mechanism is also present in flor velum yeasts during biological aging. Major alcohol content depends on the temperature and amino acid concentration among other factors. The major alcohols present in samples may be due to the Ehrlich pathway catabolic process, where amino acids act as precursors of the major alcohol synthesis [43]. Concretely, the increase in isobutanol and isoamyl alcohol content is favored by the presence of valine and leucine in the medium [44]. According to Roldán et al. (2011) [22] leucine and valine represent approximately 2% of the total amino acids present in bee pollen, thus, bee pollen addition to wines for further biological aging could increase the content of major alcohols. Despite of that, the contribution of amino acids in high doses of pollen has not produced a higher content of major alcohols. This fact could be due to a metabolic regulation since the absence of oxidized cofactors like NAD^+ would prevent the transformation of the intermediate aldehyde into major alcohols [40]. Glycerol is the second source of carbon consumed by flor velum yeasts during biological aging and nowadays it can be considered as the key compound for the flor velum metabolism analysis [35]. Glycerol consumption is not affected by the presence of bee pollen, except for the 20 g/L dose where its consumption is higher. This fact could indicate that only high doses of pollen can increase flor velum metabolism. Therefore, flor velum metabolism is not significantly affected by the presence of pollen or by its positive effect on yeast growth rate.

Regarding the results obtained from the sensory analysis (Figure 5a,b), bee pollen addition can improve the sensory attributes of biological aging in both the olfactory (dried fruit, yeast/bread, floral) and taste phase (salinity flavor and dryness sensation) when is applied in a concentration under 1 g/L. From this dose on, and specially for high doses (10–20 g/L), wines lose its characteristic sensory attributes, and acquire sensory deviations like an increase in color intensity and cereal/pollen aroma that may affect wine final quality. This color increase could be associated with the polyphenols granted by bee pollen [19] that could ease oxidation reaction development in wine in the presence of oxygen and, as a consequence, destabilize the final wine sensory quality.

4. Materials and Methods

4.1. Velum Yeast

Flor velum yeast was obtained from a biological aging system of a Sherry wine cellar and was identified as *Saccharomyces cerevisiae beticus*, a strain identified morphologically and codified as "B16" by Martínez (1995) [39]. The velum yeasts were disaggregated using a P-Selecta ultrasonic bath (Barcelona, Spain) and homogenized in a small volume of wine to be subsequently inoculated at a known concentration using a submerged protocol [33].

4.2. Fino-Sherry Wine

Fino-Sherry from Palomino Fino grapes (*Vitis vinifera*) was obtained from a 2 years old "criadera" from a "criadera y solera" aging system of the Andalusian winery "Cooperativa Unión de Viticultores Chiclaneros", Chiclana de la Frontera (Cadiz, Spain). Wine was filtered using a Whatman 0.2 μm membrane filter (Darmstad, Germany) to eliminate all suspended particles and/or microorganisms and characterized physicochemically before the biological aging trials began (initial sample).

4.3. Bee Pollen

Commercial multiflora bee pollen (Valencia, Spain) was grounded in a Vowerk's Thermomix TM31 mill (Wuppertal, Germany), sterilized under ultraviolet light for 2 h, and stored in the dark under dessicator conditions [18].

4.4. Effect on Flor Velum Yeast Growth

To study the influence of bee pollen addition on yeast development in biologically aged wines, commercial bee pollen was added to 10 mL of "Fino" sherry wine inside test tubes using six different doses: 0 (control), 0.1, and 0.25 g/L (low doses), 1 and 5 g/L (intermediate doses), and 10 and 20 g/L (high doses). Bee pollen was dissolved in wine using a P-Selecta ultrasonic bath (Barcelona, Spain). Then, yeast was inoculated at a $4.8 \times 10^3 \pm 650$ CFU/mL concentration following a submerged protocol [33]. Test tubes were stored in a temperature-controlled area at 20 ± 1 °C for 27 days using a P-Selecta lab refrigerator (Barcelona, Spain). Each bee pollen dose and control sample was conducted in triplicate (n = 3) to ensure statistical significance.

Analytical Measurements

Yeast population counts were performed using an optical Nikon microscope with 40× magnification (Tokyo, Japan) with the methylene blue staining method in a Merck Neubauer chamber (Madrid, Spain). Hydrophobicity was assessed using the toluene method [29]. Harvested flor velum yeasts cells were washed with water and suspended in a pH = 3.5 buffer, adjusting cell population as the optical density at λ = 660 nm reached 0.5. Three milliliters of this suspension was placed into another test tube and the equivalent volume of organic solvent was gently put upon the buffer. The tube was vigorously shaken using a Heidolph shaker (Schwabach, Germany). The optical density of the initial and residual buffer solution was measured. Hydrophobic degree (HD) of the yeast cell surface was calculated from the equation: HD (%) = 100 × (1−R/I); where R and I were the optical density of the initial and residual buffer solution respectively. Yeast assimilable nitrogen (YAN) was determined according to the formaldehyde method with modifications described by Aerny (1997) [45]. All the measurements were destructive analysis and were conducted in triplicate to ensure statistical significance.

4.5. *Effect of Bee Pollen on Flor Velum Metabolism*

To identify bee pollen addition influence on flor velum yeast metabolism during biological aging, oenological parameters, major volatile compounds, and organic acids were assessed in all samples. Flor velum at a concentration of $7.4 \times 10^4 \pm 114$ CFU/mL was inoculated in 200 mL of bee pollen supplemented wine inside Greiner Bio-One flasks (Kremsmünster, Germany) stored in the same conditions as the test tubes noted in Section 2.3. Under the same conditions, 200 mg/L of potassium sorbate (Sigma–Aldrich, Saint Louis, MO, United States) were added to another 200 mL flask of wine to control the ethanol amount evaporated during the biological aging process. All the experiments were conducted in triplicate. Once the experiment was finished, wines were filtered using 0.2 µm membrane filters (Darmstad, Germany) and bottled using an inert atmosphere (N_2) to avoid changes until their characterization and sensory analysis.

Analytical Measurements

Total acidity, volatile acidity, and alcohol content were determined according to the official methods of wine analysis [46]. pH determinations were carried out using a Crisson digital pH-meter (Loveland, CO, United States). Glycerol was determined using a Biosystems enzymatic kit (Barcelona, Spain) with a Thermo–Fischer UV-Vis spectrophotometer (Whaltman, MA, United States). Major volatile compounds (acetaldehyde, ethyl acetate, methanol, 1-propanol, isobutanol, and isoamyl alcohol) were determined using the method proposed by Amores-Arrocha et al. (2018) [41]. A 5 µL direct injection in an Agilent Technologies HP 5890 gas chromatograph equipped with a FID detector (Santa Clara, CA, United States) on a Sigma–Aldrich Carbowax 20M column (L 50m, ID 0.25mm, PD 0.25 µm) (Saint Louis, MO, United States). Injector temperature was 175 °C and detector temperature was 225 °C. Hydrogen was used as a carrier gas with a 1 mL/min flow. Oven temperature was set at 35 °C for the first 5 min with a 5 °C/min ramp until 100 °C were reached. 4-metil-2-propanol was used as internal standard at 783 mg/L concentration, and Sigma–Aldrich pure standard compounds (Saint Louis, MO, United States) were used to determine retention times and calibration curves. Citric, tartaric, malic, succinic, and lactic acid were assessed through ionic chromatography in a Metrohm 930 compact IC Flex ionic chromatograph equipped with a conductimetric detector on a Metrosep Organic Acids column -250/7.8- (Herisau, Switzerland). Organic acid separation was performed using as eluent a H_2SO_4 0.4 mM in a 12% acetone solution at an isocratic 0.4 mL/min flow.

4.6. Sensory Analysis

Sensory analysis was performed to find differences among the different bee pollen doses applied to wine. Three days after bottling, a 10-member expert panel in individual booths with controlled illumination performed sensory analysis. Fifty milliliters of wine were served to each taster in standard tasting glasses [47]. Each taster was given specific tasting notes to evaluate on a 10 points scale, visual (color intensity), olfactory (floral, fruity, dried fruits, etc.), and taste (saltiness, sweetness, dryness, etc.) attributes, as well as the wine general assessment. The attributes evaluated were selected according to Jackson (2002) [48].

4.7. Statistical Analysis

Means and standard deviations were calculated and significant differences were evaluated by two-way ANOVA and Bonferroni's multiple range (BSD) test with a $p < 0.05$ (GraphPad Prism version 6.01 for Windows, GraphPad Software, San Diego, CA, United States).

5. Conclusions

Bee pollen can be an effective flor velum activator in biologically aged wines when added in concentrations equal or higher than 0.25 g/L. It can reduce flor velum latency times, increasing yeast population growth rate according to the bee pollen dose. This effect may be due to the bee pollen YAN contribution to wines. During the yeast growth phase, YAN consumption per colony forming unit is significantly reduced with the pollen dose applied, possibly due to a decrease in YAN requirements by yeasts or to the fact that pollen continues releasing nitrogenized compounds throughout the biological aging process. Furthermore, bee pollen can improve and activate flor velum yeast surface hydrophobicity and its buoyancy during this aging process.

Generally, except for the 20 g/L dose, bee pollen addition does not produce any significant effect on flor velum metabolism, physicochemical parameters, organic acids content, major volatile compounds, or glycerol. The descriptive sensorial analysis determined that low bee pollen doses (0.1–0.25 g/L) increase biological aging attributes (dried fruit, yeast/bread, floral, salinity flavor, and dryness sensation), obtaining the best evaluation by the tasting panel. In this way, can be concluded that bee pollen can be a viable and natural tool to boost flor velum yeast development during biological aging and to improve the sensorial quality of final wines.

Author Contributions: P.S.-G., A.A.-A., A.J.-C., and V.P. conceived and designed the experiments. P.S.-G. performed the experiments. All authors analyzed the data and wrote the paper.

Funding: This research was funded by the AGR-203 Research Group of the University of Cadiz.

Acknowledgments: The authors thank the Andalusian cooperative-winery, "Unión de Viticultores Chiclaneros" of Chiclana de la Frontera" (Cadiz, Spain).

Conflicts of Interest: The authors declare no conflict of interest.

References

1. Martínez, P.; Codón, A.C.; Pérez, L.; Benítez, T. Physiological and molecular characterization of flor yeasts: Polymorphism of flor yeast populations. *Yeast* **1995**, *11*, 1399–1411. [CrossRef] [PubMed]
2. Pozo-Bayón, M.A.; Moreno-Arribas, M.V. Sherry Wines. In *Advances in Food and Nutrition Research*, 1st ed.; Jackson, R.S., Ed.; Academic Press, Elsevier Ltd.: London, UK, 2011; Volume 63, pp. 17–40.
3. Moreno-García, J.; Mauricio, J.C.; Moreno, J.; García-Martínez, T. Stress responsive proteins of a flor yeast strain during the early stages of biofilm formation. *Process Biochem.* **2016**, *51*, 578–588. [CrossRef]
4. Alexandre, H. Flor yeasts of Saccharomyces cerevisiae-Their ecology, genetics and metabolism. *Int. J. Food Microbiol.* **2013**, *167*, 269–275. [CrossRef]
5. Esteve-Zarzoso, B.; Fernández-Espinar, M.T.; Querol, A. Authentication and identification of Saccharomyces cerevisiae "flor" yeast races involved in sherry ageing. *Antonie Van Leeuwenhoek* **2004**, *85*, 151–158. [CrossRef]

6. Berlanga, T.M.; Peinado, R.; Millán, C.; Mauricio, J.C.; Ortega, J.M. Influence of Blending on the Content of Different Compounds in the Biological Aging of Sherry Dry Wines. *J. Agric. Food. Chem.* **2004**, *52*, 2577–2581. [CrossRef]
7. Peinado, R.A.; Mauricio, J.C. Biologically aged wines. In *Wine Chemistry and Biochemistry*, 1st ed.; Moreno-Arribas, M.C., Polo, C., Eds.; Springer-Verlag: New York, NY, USA, 2009; pp. 81–101.
8. Moreno-García, J.; García-Martínez, T.; Moreno, J.; Mauricio, J.C. Proteins involved in flor yeast carbon metabolism under biofilm formation conditions. *Food Microbiol.* **2015**, *46*, 25–33. [CrossRef]
9. Ibeas, J.I.; Lozano, I.; Perdigones, F.; Jiménez, J. Effects of Ethanol and Temperature on the Biological Aging of Sherry Wines. *Am. J. Enol. Viticult.* **1997**, *48*, 71–74.
10. Berlanga, T.M.; Millán, C.; Mauricio, J.C.; Ortega, J.M. Influence of nitrogen on the biological aging of Sherry wine. *J. Sci. Food Agric.* **2006**, *86*, 2113–2118. [CrossRef]
11. Arias-Gil, M.; Garde-Cerdán, T.; Ancín-Azpilicueta, C. Influence of addition of ammonium and different amino acid concentrations on nitrogen metabolism in spontaneous must fermentation. *Food Chem.* **2007**, *103*, 1312–1318. [CrossRef]
12. Cooper, T.G. Nitrogen metabolism in Saccharomyces cerevisiae. In *The Molecular Biology of the yeast Saccharomyces. Metabolism and Gene Expression*, 2nd ed.; Strathern, J.N., Jones, E.W., Broach, J.R., Eds.; Cold Spring Harbor Laboratory: Cold Spring Harbor, NY, USA, 1982; pp. 39–99.
13. Large, P.J. Degradation of organic nitrogen compounds by yeasts. *Yeasts* **1986**, *2*, 1–34. [CrossRef]
14. Suarez-Lepe, J.A.; Leal, Í. *Microbiología Enológica. Fundamentos de Vinificación*, 3rd ed.; Editorial Mundi-Prensa: Barcelona, Spain, 1990; pp. 110–111.
15. Mauricio, J.C.; Moreno, J.J.; Ortega, J.M. In Vitro Specific Activities of Alcohol and Aldehyde Dehydrogenases from Two Flor Yeasts during Controlled Wine Aging. *J. Agric. Food Chem.* **1997**, *45*, 1967–1971. [CrossRef]
16. Mauricio, J.C.; Valero, E.; Millán, C. Changes in Nitrogen Compounds in Must and Wine during Fermentation and Biological Aging by Flor Yeasts. *J. Agric. Food Chem.* **2001**, *49*, 3310–3315. [CrossRef]
17. Cortes, M.; Moreno, J.J.; Zea, L.; Moyano, L.; Medina, M. Response of the aroma fraction in Sherry wines subjected to accelerated biological aging. *J. Agric. Food Chem.* **1999**, *47*, 3297–3302. [CrossRef] [PubMed]
18. Amores-Arrocha, A.; Roldán, A.; Jiménez-Cantizano, A.; Caro, I.; Palacios, V. Effect on White Grape Must of Multiflora Bee Pollen Addition during the Alcoholic Fermentation Process. *Molecules* **2018**, *23*, 1321. [CrossRef]
19. Almeida-Muradian, L.B.; Pamplona, L.C.; Coimbra, S.; Barth, O.M. Chemical composition and botanical evaluation of dried bee pollen pellets. *J. Food Compos. Anal.* **2005**, *18*, 105–111. [CrossRef]
20. Human, H.; Nicolson, S.W. Nutritional content of fresh, bee-collected and stored pollen of Aloe greatheadii var. davyana (Asphodelaceae). *Phytochemistry* **2006**, *67*, 1486–1492. [CrossRef] [PubMed]
21. Campos, M.G.R.; Bogdanov, S.; Almeida-Muradian, L.B.; Szczesna, T.; Mancebo, Y.; Frigerio, C.; Ferreira, F. Pollen composition and standardization of analytical methods. *J. Apic. Res.* **2008**, *47*, 154–161. [CrossRef]
22. Roldán, A.; Van Muiswinkel, G.C.J.; Lasanta, C.; Palacios, V.; Caro, I. Influence of pollen addition on mead elaboration: Physicochemical and sensory characteristics. *Food Chem.* **2011**, *126*, 574–582. [CrossRef]
23. Henschke, P.A.; Jiranek, V. Metabolism of nitrogen compounds. In *Wine Microbiology and Biotechnology*; Fleet, G.H., Ed.; Harwood Academic: Lausanne, Switzerland, 1993; pp. 77–164.
24. Ribéreau-Gayon, P.; Dubourdieu, D.; Donéche, B.; Lonvaud, A.; Glories, Y.; Maugean, A. *Tratado de Enología: Microbiología del vino. Vinificaciones*, 2nd ed.; Hemisferio Sur: Buenos Aires, Argentina, 2003.
25. Dos Santos, A.; Feuillat, M.; Charpentier, C. Flor yeast metabolism in a model system similar to cellar ageing of the French "Vin Jaune": Evolution of some by-products, nitrogen compounds and polysaccharides. *VITIS* **2000**, *39*, 129–134.
26. Sroka, P.; Tuszýnski, T. Changes in organic acids contents during mead wort fermentation. *Food Chem.* **2007**, *104*, 1250–1257. [CrossRef]
27. Xu, X.; Sun, L.; Dong, J.; Zhang, H. Breaking the cells or rape bee pollen and consecutive extraction of functional oil with supercritical carbon dioxide. *Innov. Food Sci. Emerg. Technol.* **2009**, *10*, 42–46. [CrossRef]
28. Pérez-Serradilla, J.A.; Luque de Castro, M.D. Role of lees in wine production: A review. *Food Chem.* **2008**, *111*, 447–456. [CrossRef]
29. Iimura, Y.; Hara, S.; Otsuka, K. Cell Surface Hydrophobicity as a Pellicle Formation Factor in Film Strain of Saccharomyces. *Agric. Biol. Chem.* **1980**, *1369*, 1215–1222.

30. Aguilera, F.; Valero, E.; Millán, C.; Mauricio, J.C.; Ortega, J.C. Cellular fatty acid composition of two physiological races of Saccharomyces cerevisiae during fermentation and flor veil formation in biological aging of fine wines. *Belgian J. Brewing Biotechnol.* **1997**, *2*, 39–42.
31. Amores-Arrocha, A. Aplicación del polen de abeja como activador en el proceso de fermentación alcohólica. Ph.D. Thesis, Facultad de Ciencias, Universidad de Cádiz, Cádiz, Spain, July 2018.
32. Berlanga, T.M.; Atanasio, C.; Mauricio, J.C.; Ortega, J.M. Influence of Aeration on the Physiological Activity of Flor Yeasts. *J. Agric. Food Chem.* **2001**, *49*, 3378–3384. [CrossRef]
33. Pérez, L.; Valcárcel, M.J.; González, P.; Domecq, B. Influence of Botrytis Infection of the Grapes on the Biological Aging Process of Fino Sherry. *Am. J. Enol. Vitic.* **1991**, *42*, 58–62.
34. Roldán, A.M.; Lloret, I.; Palacios, V. Use of a submerged yeast culture and lysozyme for the treatment of bacterial contamination during biological aging of sherry wines. *Food Control* **2017**, *71*, 42–49. [CrossRef]
35. Zea, L.; Serratosa, P.; Julieta, M. Acetaldehyde as Key Compound for the Authenticity of Sherry Wines: A Study Covering 5 Decades. *Compr. Rev. Food Sci. Food Saf.* **2015**, *14*, 681–693. [CrossRef]
36. Zoecklein, B.W.; Fugelsang, K.C.; Gump, B.H.; Nury, F.S. Metals, Cations and Anions. In *Wine Analysis and Production*, 1st ed.; Springer: New York, NY, USA, 1999; pp. 199–208.
37. Gerbaux, V.; Villa, A.; Monamy, C.; Bertrand, A. Use of Lysozyme to Inhibit Malolactic Fermentation and to Stabilize Wine After Malolactic Fermentation. *Am. J. Enol. Vitic.* **1997**, *48*, 49–54.
38. Martínez de la Ossa, E.; Pérez, L.; Caro, I. Variations of the major volatiles through aging of sherry. *Am. J. Enol. Vitic.* **1987**, *38*, 293–297.
39. Martínez, P. Evolución y caracterización de las poblaciones de levaduras responsables de la crianza biológica del vino de Jerez. Ph.D. Thesis, Universidad de Sevilla, Seville, Spain, February 1995.
40. Swiegers, J.H.; Bartowsky, E.; Pretorius, I.S. Yeast and bacterial modulation of wine aroma and flavor. *Aust. J. Grape Wine. R.* **2005**, *11*, 139–173. [CrossRef]
41. Mauricio, J.C.; Moreno, J.J.; Valero, E.M.; Zea, L.; Medina, M.; Ortega, J.M. Ester Formation and Specific Activities of in Vitro Alcohol Acetyltransferase and Esterase by Saccharomyces cerevisiae during Grape Must Fermentation. *J. Agric. Food Chem.* **1993**, *41*, 2086–2091. [CrossRef]
42. Amores-Arrocha, A.; Roldán, A.; Jiménez-Cantizano, A.; Caro, I.; Palacios, V. Evaluation of the use of multiflora bee pollen on the volatile compounds and sensorial profile of Palomino fino and Riesling white young wines. *Food Res. Int.* **2018**, *105*, 197–209. [CrossRef]
43. Styger, G.; Prior, B.; Bauer, F.F. Wine flavor and aroma. *J. Ind. Microbiol. Biotechnol.* **2011**, *38*, 1145–1159. [CrossRef]
44. Boulton, R.B.; Singleton, V.L.; Bisson, L.F.; Kunkee, R.E. Yeast and Biochemistry of Ethanol Fermentation. Boulton. In *Principles and Practices of Winemaking*, 1st ed.; Springer: New York, NY, USA, 1999; pp. 102–192.
45. Aerny, J. Composés azotés des moûts et des vins. *Rev. Suisse Vitic. Arboric. Hortic.* **1997**, *28*, 161–168.
46. OIV Office international de la Vigne et du Vin. *Recuéil des méthodes internationales d'analyse des vins et des moûts*; Edition Officielle: Paris, France, 2014.
47. ISO 3591. *Sensory Analysis—Apparatus—Wines-Tasting Glass*; International Organization for Standardization: Genève, Switzerland, 1997.
48. Jackson, R.S. Wine Tasting: A Professional Handbook. Academic Press: London, UK, 2009.

Sample Availability: Samples of the compounds are not available.

molecules

MDPI

Article

Applying Nanoparticle Tracking Analysis to Characterize the Polydispersity of Aggregates Resulting from Tannin–Polysaccharide Interactions in Wine-Like Media

Sijing Li [1,2,†], Kerry L. Wilkinson [1,2], Agnieszka Mierczynska-Vasilev [3] and Keren A. Bindon [3,*]

1 School of Agriculture, Food and Wine, The University of Adelaide, PMB 1, Glen Osmond, SA 5064, Australia; sli@csu.edu.au (S.L.); kerry.wilkinson@adelaide.edu.au (K.L.W.)
2 The Australian Research Council Training Centre for Innovative Wine Production, PMB 1, Glen Osmond, SA 5064, Australia
3 The Australian Wine Research Institute, PO Box 197, Glen Osmond, SA 5064, Australia; Agnieszka.Mierczynska-Vasilev@awri.com.au
* Correspondence: keren.bindon@awri.com.au; Tel.: +61-8-8313-6601
† Current address: National Wine and Grape Industry Centre, Charles Sturt University, Wagga Wagga, NSW 2678, Australia

Academic Editors: Rosa Maria de Sá Perestrelo and José Sousa Câmara
Received: 26 April 2019; Accepted: 29 May 2019; Published: 3 June 2019

Abstract: Interactions between grape seed tannin and either a mannoprotein or an arabinogalactan in model wine solutions of different ethanol concentrations were characterized with nanoparticle tracking analysis (NTA), UV-visible spectroscopy and dynamic light scattering (DLS). NTA results reflected a shift in particle size distribution due to aggregation. Furthermore, the light scattering intensity of each tracked particle measured by NTA demonstrated the presence of aggregates, even when a shift in particle size was not apparent. Mannoprotein and arabinogalactan behaved differently when combined with seed tannin. Mannoprotein formed large, highly light-scattering aggregates, while arabinogalactan exhibited only weak interactions with seed tannin. A 3% difference in alcohol concentration of the model solution (12 vs. 15% *v/v*) was sufficient to affect the interactions between mannoprotein and tannin when the tannin concentration was high. In summary, this study showed that NTA is a promising tool for measuring polydisperse samples of grape and wine macromolecules, and their aggregates under wine-like conditions. The implications for wine colloidal properties are discussed based on these results.

Keywords: mannoprotein; arabinogalactan; seed tannin; aggregation; nanoparticle tracking analysis

1. Introduction

Condensed tannins are among the most abundant macromolecules in red wine. They are extracted from grape skins and seeds during maceration and predominantly consist of condensed polymers of flavan-3-ols, at concentrations up to 4 mg/mL [1]. Under wine-like conditions, condensed tannins can aggregate and form colloidal dispersions, with hydrodynamic diameters in the magnitude of a few hundred to over a thousand nanometers [2,3]. The mean particle size of red wine colloids were shown to increase after 18 months of bottle aging, which was suspected to be responsible for the decrease in tannin concentration and molecular size observed in wines after aging [4]. These colloids may continue to aggregate and eventually precipitate and form sediment in wine bottles [5]. Thus, colloidal dispersion in wine is closely linked to wine mouthfeel perceptions and color stabilization, and warrants continuing research.

Polysaccharides in wine have been found to associate with condensed tannin non-covalently, through hydrogen bonding and hydrophobic interactions [6] and affect the colloid size evolution [7,8]. Polysaccharides are present in wine from 0.2 to 1.5 mg/mL, and consist predominately of neutral polysaccharides, which are mainly arabinogalactan-protein derived from the grape cell wall and mannoprotein derived from the yeast involved in fermentation [9]. Previous studies observed that polysaccharides mediate interactions between tannins and proteins [10–12], conferring impacts on wine mouthfeel [13], color stabilization [14] and fining (removal) of phenolic compounds [15]. For these reasons, polysaccharides have been used by the wine industry to improve wine composition and organoleptic characters. In Australia, two types of commercially manufactured polysaccharide additives are permitted in wine production: yeast mannoprotein and gum arabic (arabinogalactan) [16]. Supplementing wine with commercial polysaccharide products is likely to impact the colloidal stability of wine. It is therefore important to characterize the impact on colloidal dispersion in wine following the addition.

Characterizing interactions between grape- and wine-derived polysaccharides and tannins poses unique challenges, since both materials are very polydisperse. Grape skin tannins can be comprised of three to 83 flavan-3-ol subunits, while seed tannins are reported to have 2 to 16 subunits [1]. Wine polysaccharides are also heterogeneous, with molecular distribution reported to be between 5 and 800 kDa [9]. In addition, these macromolecules tend to interact and aggregate under wine conditions. Different fractions of macromolecules isolated from Pinot Noir wines, including tannins, polysaccharides and proteins, were shown to be highly polydisperse, with particle size distributions ranging from 20 to 500 nm [17]. Moreover, the properties of the dispersant have a significant impact on the macromolecular interaction, e.g., pH, ethanol concentration and ionic strength [8]. Thus, investigations into these interactions require non-invasive techniques, so as not to disrupt the non-covalent associations between particles, and at the same time, detect aggregate formation in a wine-like medium. Methods that have been employed to study polysaccharide and condensed tannin interactions include HPLC, nephelometry, Saturation-Transfer Difference-NMR, UV-Visible spectroscopy, dynamic light scattering, isothermal titration calorimetry, small-angle X-ray scattering and transmission electron microscopy [8,18–22].

Nanoparticle particle tracking analysis (NTA) is a relatively new technology (first commercialized in 2006) in nanoparticle characterization. It tracks the Brownian motion of individually recognized particles to deduce their hydrodynamic diameter. Since particles are individually analyzed, the size distribution in polydisperse samples is less skewed towards the large particles, hence giving more accurate results [23]. NTA has been applied to many food matrices and can handle non-aggressive solvents such as hydroalcoholic solutions [24]. Thus, NTA presents a promising technique to investigate the polydisperse colloidal dispersion in wine. In the current study, we aimed to investigate to what extent applying commercial polysaccharide additives would affect the colloidal state of wine, and by inference, the composition and organoleptic characters of wine. To this end, two polysaccharides purified from two commercial wine additives and a purified grape seed tannin fraction were combined in model wine solutions at two wine-like alcohol concentrations. To date, and to our knowledge, NTA has not been used to study tannin and polysaccharide interactions. Thus, the current study also aimed to evaluate the suitability of NTA for this type of investigation, corroborated by other techniques that have been successfully applied in this field.

2. Results and Discussion

2.1. Molecular Weight and Particle Size Distribution of Seed Tannin, Mannoprotein and Gum Arabic

Size exclusion chromatography showed that seed tannin (ST), mannoprotein (MP) and gum arabic (AG) had moderate polydispersity indices, from 1.8 to 2.3 (Table 1). However, the molecular weight ranges were substantially different. The molecular weight of ST ranged from 0.5 to 6 kg/mol, with a mean of 1.9 kg/mol, which approximated the degree of polymerization (DP) 6 [25]. In contrast, the two polysaccharides had much higher molecular weight ranges, 10–98 kg/mol and 48–322 kg/mol for MP and AG respectively, which were within the range that is typically observed for wine polysaccharides [9].

Table 1. Mean and range of molecular weight (M_r) and polydispersity index (PdI) determined by size exclusion chromatography.

Material	Mean M_r (kg/mol)	PdI	M_r10 (kg/mol)	M_r90 (kg/mol)
Mannoprotein (MP)	33.6 ± 1.9	2.25 ± 0.07	9.6 ± 0.3	97.9 ± 8.8
Gum arabic (AG)	132.5 ± 1.0	1.75 ± 0.01	47.6 ± 0.4	321.6 ± 1.4
Seed tannin (ST)	1.92 ± 0.02	2.14 ± 0.01	0.5 ± 0.01	6.0 ± 0.09

All values are means of 3 measurements ± standard error. M_r10 and M_r90 stand for molecular weight at the 10th and 90th percentiles of elution respectively.

The polydispersity index (PdI) of MP and AG determined by DLS were 0.5 and 0.9 respectively, while the cumulant fit error was higher than 0.005 for ST, indicating poor data quality or high sample polydispersity [26]. As such, the average particle size derived from cumulant analysis are not reported. Instead, intensity based particle size distributions, which are appropriate for polydisperse samples, are reported in Figure 1. DLS detected a peak between 10 and 60 nm for both MP and AG, as well as between 1 and 10 nm for ST, which were not detected by NTA. Both methods detected particles above 60 nm. This was possibly caused by the different detection limits of the two methods; for biological particles, the lower detection threshold for NTA is 60 nm ± 30% [27], while it is 1 nm for DLS [23]. The broad particle size distribution measured by DLS indicated that a range of aggregate sizes existed in the solutions. For MP and AG (Figure 1A,B), the intensity distribution showed that the particles of lower size were the primary component, considering intensity of light scattered by particles is proportional to the sixth power of its diameter. AGs isolated from commercial gum arabic generally have gyration radii between 20 and 30 nm [28], and tend to form particulate aggregates (≈100 nm) at concentrations as low as 0.1% in aqueous solution [29], where size increases with AG concentration. Commercial MP products also form aggregates when reconstituted in model wine solutions [30]. It was therefore considered that the observed aggregates >60 nm presented stable colloidal suspensions rather than insoluble material, which would be expected to have been removed during the centrifugation step [7].

The two groups of ST, at 4.1 and 256.6 nm (Figure 1C) could be detected by DLS, which were comparable to respective gyration radii of molecularly dissolved or aggregated grape seed tannins measured by small angle neutron scattering [31]. Using NTA, the major group of larger aggregates detected had a narrower distribution than determined by DLS, with a mean particle size of 91 nm. According to Zanchi and colleagues [31], 2% of native seed tannin loses solubility below 60% *v/v* ethanol, and represents a group of hydrophobic, oxidized tannins. Between 60% *v/v* ethanol and wine-like ethanol concentration (12% *v/v*) the bulk of tannin molecules are expected to remain in solution, while a sub-group (33%) forms metastable colloidal dispersions. It would be expected that the larger aggregates detected by DLS and NTA represented colloidally dispersed particles. According to Zanchi et al. [31,32], tannins which form aggregates between 12 and 60% *v/v* ethanol are preferentially solvated in ethanol, and that the proportion falling into this class may be enhanced by oxidation. The phloroglucinolysis conversion yield for ST was 67% (Supplementary Table S1), and it might be hypothesized that the uncharacterized portion indicates the presence of oxidized tannins in ST. This conversion yield is typical of grape seeds, which are thought to undergo oxidation in situ during ripening [25], but which may also potentially occur during the extraction process. In the literature, conversion yields of seed tannins were from 60 to 86% [33,34]. For comparison, a Tannat seed tannin of high conversion yield 86% [35] was compared with the ST used in this study, and produced the same particle distribution profile (data not shown). It would therefore be expected that any insoluble material would have been removed with the centrifugation step, as observed by Zanchi et al. [32] and that a portion of the oxidized tannin in the ST sample remained as a metastable colloidal dispersion in ethanol solution, while the remainder of the tannin was in solution (≤4 nm).

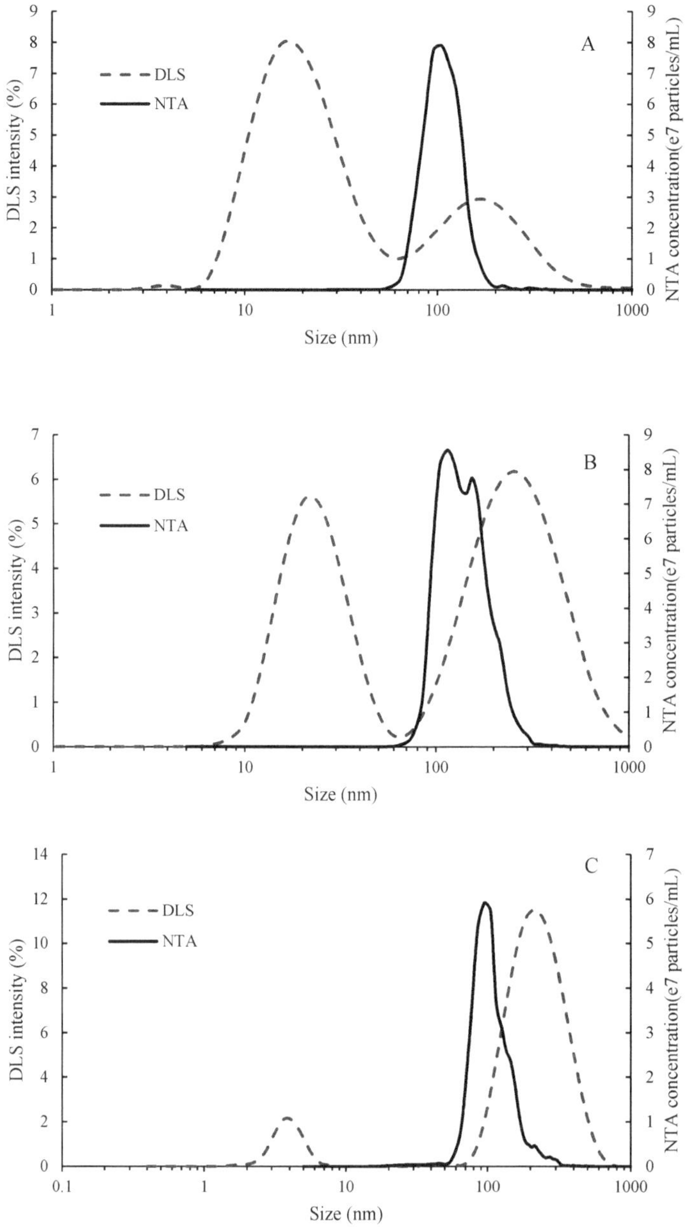

Figure 1. Particle size distribution of (**A**) MP, (**B**) AG and (**C**) ST, measured by dynamic light scattering (DLS) and nanoparticle tracking analysis (NTA).

2.2. Interactions between Polysaccharides and Tannins Characterized by UV-Visible Spectrometry

Formation of aggregates between neutral polysaccharides and ST at a range of concentrations (0.065–5 mg/mL) were determined by measuring their absorbance at 650 nm. Since neither of these substances absorb light at this wavelength, the absorbance value is dominated by the light scattering intensity of particles, and therefore can serve as an indication of aggregate formation [6,20]. An initial absorbance in MP and AG solutions in the absence of ST indicated aggregates were naturally present in these solutions (Figure 2). For ST, a sharp increase in 650 nm absorbance was observed at lower ST concentrations, i.e., up to 1.25 mg/mL, followed by a steadier rise to 5 mg/mL, in both model wine solutions. This was likely due to decreased solubility at increasing concentrations. Absorbance of the mixture of ST and AG followed an identical trend to that of ST. In contrast to AG, the combination of MP and ST did not result in increases in absorbance at the lower ST concentrations. However, the 650 nm absorbance increased substantially in the MP and ST mixtures at the higher tannin concentrations of 2.5 and 5 g/mL in 12% model wine, indicating formation of highly scattering large particles (Figure 2A). Interestingly, in 15% model wine, the absorbance of the MP and ST combination increased evenly across the tannin concentration gradient. Strong increases in absorbance at 650 nm have been reported between a protein-rich arabinogalactan-protein (AGP) and procyanidins (DP 30) at high concentrations, although the absorbance reported was much higher than that found in the current study [6]. The UV-visible spectrometry measurement of MP, AG and ST has since been replicated in our lab (data not shown), and an analogous trend to the current study was found. The increase in absorbance at 650 nm for ST and MP in 12% model wine was further explored with NTA and DLS.

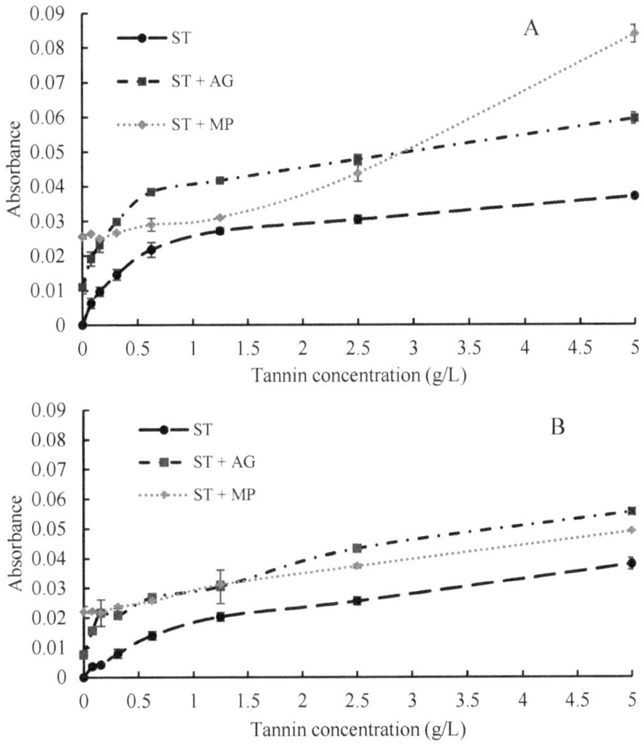

Figure 2. Absorbance (650 nm) of seed tannin from 0 to 5 mg/mL, with or without addition of polysaccharides in (**A**) 12% ethanol model wine and (**B**) 15% ethanol model wine.

Absorbance at 280 nm was recorded in order to assess the impact of polysaccharide addition on phenolic content (retention or precipitation from solution). The absorbance values at 280 nm are reported in Supplementary Table S2. Regression analysis showed that the absorbance at 280 nm increased linearly ($R^2 > 0.99$) with tannin concentrations, and was not affected by centrifugation or the alcohol concentration of the model wine. Although statistical analyses showed some differences in the absorbance at 280 nm between ST and the combination of ST and polysaccharides at certain tannin concentrations, there was a lack of consistency in the differences and no general trend could be attributed to the tannin concentration, polysaccharide type, centrifugation or ethanol concentration in the model wine (Supplementary Table S2). It was therefore assumed that the addition of polysaccharide did not influence the total phenolic concentration under the conditions used in the current study. No loss at 280 nm absorbance was observed in the ST and polysaccharide mixtures before or after mild centrifugation, indicating that centrifuging did not remove aggregates formed between tannin and polysaccharide. This was consistent with the report that aggregates formed between tannin and polysaccharide have low density and do not sediment with ultracentrifugation [19].

2.3. Binding Experiment Characterized by NTA

Based on the UV-vis spectroscopy results, two ST concentration points were further characterized by NTA and DLS: 1.25 and 5 mg/mL ST, combined with 0.5 mg/mL of either MP or AG, in both 12 and 15% model wine solutions.

Number-weighted size distributions of particles of ST, MP, AG and their mixtures, were determined by NTA and compared (Figures 3 and 4). Notably, the absolute concentrations (number of particles/mL) between samples were not compared in this instance because the camera settings and detection threshold were optimized for each sample and may have therefore affected particle recognition and count for each size class (and thus affect particle concentration). As a result, comparison of the distribution only aimed to identify shifts in particle sizes, in order to infer the formation of aggregates. NTA also determined particle size at the 10th, 50th and 90th percentiles of the distribution, as well as an overall mean. These numerical data were also reported for ease of comparison (Table 2).

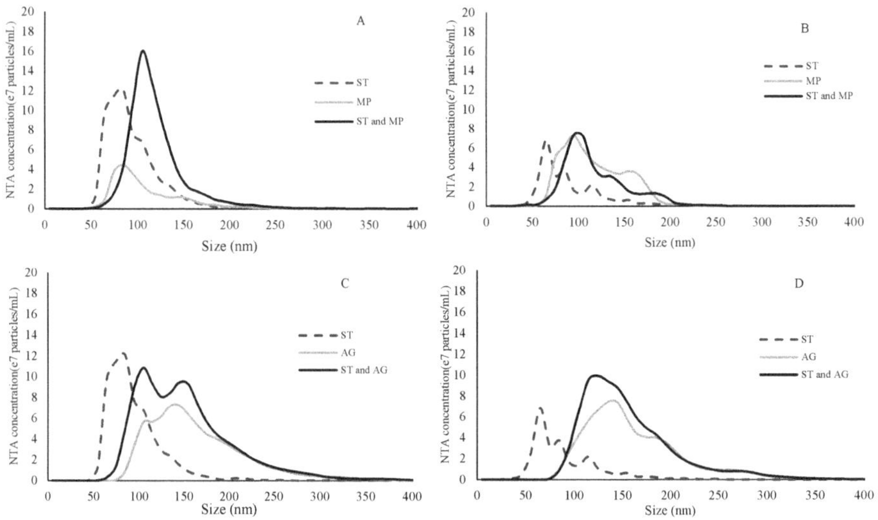

Figure 3. Size distribution of binding experiments between 1.25 mg/mL tannin and 0.5 mg/mL polysaccharides determined by nanoparticle tracking analysis. The curves were an average of 15 measurements. (**A**) ST and MP in 12% ethanol model wine; (**B**) ST and MP in 15% ethanol model wine; (**C**) ST and AG in 12% ethanol model wine; and (**D**) ST and AG in 15% ethanol model wine. All ST-containing solutions were diluted 1:10 with model wine prior to analysis.

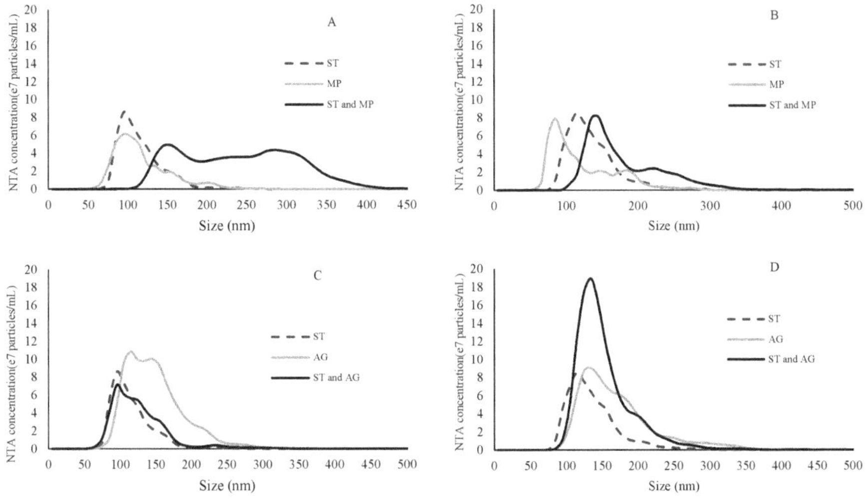

Figure 4. Size distribution of binding experiments between 5 mg/mL tannin and 0.5 mg/mL polysaccharides determined by nanoparticle tracking analysis. The curves are from an average of 15 measurements. (**A**) ST and MP in 12% ethanol model wine; (**B**) ST and MP in 15% ethanol model wine; (**C**) ST and AG in 12% ethanol model wine; and (**D**) ST and AG in 15% ethanol model wine. All ST-containing solutions were diluted 1:40 with model wine prior to analysis.

At 1.25 mg/mL, ST particles were smaller than either of the polysaccharides, and when ST was combined with either polysaccharide type, the size distribution of the mixture shifted towards a higher average (Figure 3). The overall particle size of the ST and MP combination was slightly higher than that of MP alone in 12% model wine, but not in 15% model wine (Table 2). On the other hand, at both ethanol levels, the ST and AG combination or AG alone had almost identical size distributions, although AG alone had a slightly higher mean size than the mixture. At 5 mg/mL, ST formed larger particles than at 1.25 mg/mL, which were comparable or slightly larger than MP, but still smaller than AG (Table 2). The trend of particle size evolution between ST, AG and their mixtures observed at lower tannin concentrations generally held true in samples containing 5 mg/mL ST. However, obvious formation of aggregates between ST and MP could be detected at this tannin concentration. In particular, in 12% model wine, very large particles of between 250 and 400 nm could be found (Figure 4A). In general, the aggregate formation between AG and ST was relatively unaffected by either tannin or alcohol concentrations. In contrast, MP formed significantly larger aggregates at higher tannin concentrations which were further promoted as alcohol was lowered.

Table 2. Particle size of seed tannin (ST) combined with commercial mannoprotein (MP) or gum arabic (AG) at 0.5 mg/mL measured by nanoparticle tracking analysis.

ST [a] (mg/mL)	Model Wine [b]	Treatment	Percentile and Mean Particle Size by NTA (nm)				Treatment	Percentile and Mean Particle Size by NTA (nm)			
			D10	D50	D90	Mean		D10	D50	D90	Mean
1.25 [c]	12%	ST	56.4 ± 0.8	79.1 ± 1.3	121.9 ± 2.5	91.4 ± 1.4	ST	56.4 ± 0.8	79.1 ± 1.3	121.9 ± 2.5	91.4 ± 1.4
		MP	62.8 ± 0.6	87.7 ± 0.9	158.3 ± 5.0	105.7 ± 1.0	AG	96.0 ± 0.9	143.5 ± 0.8	226.9 ± 2.9	160.1 ± 1.0
		ST + MP	80.2 ± 0.8	104.0 ± 0.9	152.5 ± 3.3	116.5 ± 1.0	ST + AG	85.8 ± 1.3	133.1 ± 1.2	214.1 ± 2.8	147.6 ± 1.4
	15%	ST	48.3 ± 0.9	71.4 ± 1.9	124.0 ± 4.9	85.7 ± 1.8	ST	48.3 ± 0.9	71.4 ± 1.9	124.0 ± 4.9	85.7 ± 1.8
		MP	68.7 ± 1.0	101.1 ± 1.2	155.6 ± 1.3	113.2 ± 0.8	AG	98.2 ± 1.4	141.0 ± 2.1	223.1 ± 4.3	157.5 ± 1.7
		ST + MP	73.3 ± 0.9	99.0 ± 0.8	160.1 ± 3.3	113.9 ± 1.0	ST + AG	97.3 ± 0.9	134.5 ± 1.0	209.9 ± 3.4	151.0 ± 0.9
5 [d]	12%	ST	78.3 ± 1.2	99.6 ± 1.2	147.3 ± 4.7	112.5 ± 1.3	ST	78.3 ± 1.2	99.6 ± 1.2	147.3 ± 4.7	112.5 ± 1.3
		MP	71.5 ± 1.3	100.8 ± 2.1	160.9 ± 4.9	113.4 ± 2.0	AG	93.8 ± 0.7	132.6 ± 1.8	193.4 ± 4.1	144.2 ± 1.9
		ST + MP	134.5 ± 2.8	229.2 ± 3.5	316.2 ± 3.8	231.2 ± 3.1	ST + AG	77.3 ± 1.0	108.6 ± 1.5	160.9 ± 6.0	121.4 ± 2.0
	15%	ST	90.5 ± 1.3	119.7± 1.7	170.4 ± 2.9	131.4 ± 1.5	ST	90.5 ± 1.3	119.7± 1.7	170.4 ± 2.9	131.4 ± 1.5
		MP	68.0 ± 1.0	100.2 ± 4.2	184.1 ± 7.7	120.7 ± 3.2	AG	107.2 ± 1.1	148.2 ± 2.2	237.9 ± 5.6	167.1 ± 2.2
		ST + MP	117.9 ± 1.5	153.3 ± 2.5	249.5 ± 6.2	176.1 ± 2.5	ST + AG	105.1 ± 1.0	132.9 ± 1.0	192.8 ± 2.6	147.3 ± 1.3

Values are means of 15 measurements ± standard error. [a] Seed tannin concentration in the solutions. [b] Ethanol concentrations in the model wine solutions (*v*/*v*). [c] All solutions containing seed tannin were diluted 1:10 with model wine prior to analysis. [d] All solutions containing seed tannin were diluted 1:40 with model wine prior to analysis.

NTA also provided light-scattering intensity data for each tracked particle. Since camera settings affect the light scattering intensity recorded for samples, four sets of experiments where similar camera settings (Supplementary Table S3) were applied to all samples, were chosen for presentation in Figure 5. Some low-light scattering particles were detected in MP and ST samples individually, but only higher-light scattering particles were detected in the mixtures, especially when ST concentration was high (Figure 5A,B), indicating aggregation (interactions) between ST and MP. In contrast, no clear difference could be seen in the light scattering intensity of the mixture compared to AG and ST individually, with respect to ST and ethanol concentration (Figure 5C,D). It is interesting to note that, despite of increased particle light scattering intensity in the mixture of 1.25 mg/mL ST and MP in 12% ethanol model wine (Figure 5A), the particle size distribution shift was not apparent (Figure 3A and Table 2). Potentially, in this mixture, small oligomeric aggregates formed between the tannin and polysaccharide, but these were not bridged together to form larger aggregates due either to low tannin concentration or a low degree of polymerization [19,36]. However, further investigation was required to determine whether the disappearance of low-light scattering particles in ST and MP mixtures was truly due to tannin and polysaccharide aggregation or the limit of NTA measurement. It needed to be ascertained whether a mixture of two groups of particles of distinct yet similar sizes could be discriminated by NTA; i.e., if particles of larger size could potentially dominate the measurements. To further qualify this possibility, a small amount of 100 nm polystyrene beads was mixed with AG in 12% model wine solution and the size distribution was measured by NTA (Figure 6). In the size distribution profile of the mixture, both a distinctive peak of approximately 100 nm and a broader shoulder between 150 and 300 nm could be identified, representing the beads and the AG particles, respectively. Since the camera settings were similar between mixtures and their components (i.e., ST and polysaccharides), we concluded that if the smaller ST and MP particles were present in substantial quantities in the mixture, they should not have been entirely obscured by the larger species, and if present, should therefore have been detected.

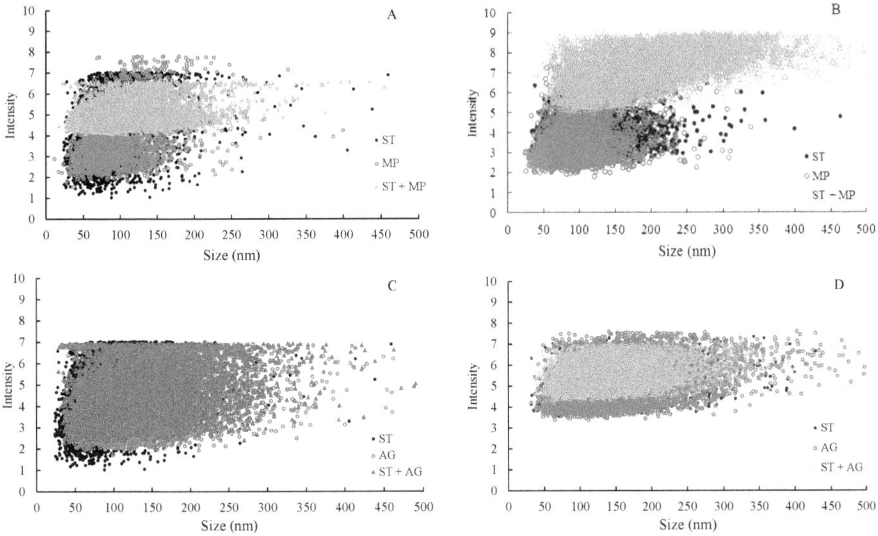

Figure 5. Size vs. light scattering intensity (arbitrary unit) for each tracked particle in nanoparticle tracking analysis. Only 1/5 of all tracked particles were included in the figures for clarity. (**A**) 1.25 mg/mL ST and 0.5 mg/mL MP in 12% ethanol model wine; (**B**) 5 mg/mL ST and 0.5 g/mL MP in 15% ethanol model wine; (**C**) 1.25 mg/mL ST and 0.5 mg/mL AG in 12% ethanol model wine; and (**D**) 5 mg/mL ST and 0.5 mg/mL AG in 15% ethanol model wine. ST solutions containing 1.25 mg/mL or 5 mg/mL respectively were diluted 1:10 or 1:40 with the corresponding model wine prior to analysis.

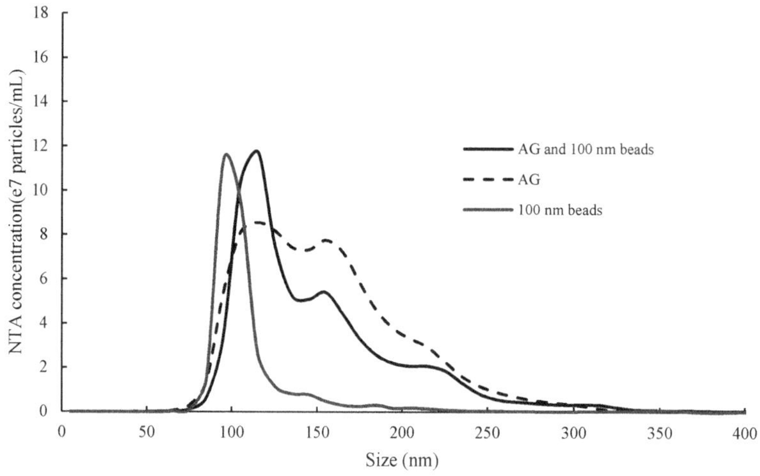

Figure 6. Size distribution of 100 nm polystyrene beads, AG and their mixture, determined by nanoparticle tracking analysis.

The different behaviours between MP and AG towards ST was also explored with DLS. At both 0.125 and 0.5 mg/mL ST concentrations, the MP and ST combination resulted in a significantly higher light scattering intensity than was observed for the AG and ST combination (Supplementary Figure S1). In particular, at 0.5 mg/mL ST in 12% model wine solution, the light scattering intensity of the MP and ST combination was 7 times higher than that of the ST and AG combination. The DLS results confirmed those measured by NTA. Furthermore, DLS detected multiple particle size groups (peaks) and high PdI values in the AG and ST mixture (Supplementary Table S4), which were very similar in size to those observed in AG and ST separately (Figure 1). Conversely, the MP and ST combination showed only one apparent size group (PdI = 0.2), irrespective of ST and ethanol concentration. These results, together with results from NTA, strongly suggested that MP and ST formed aggregates under the current experimental conditions, while AG and ST had very weak interactions and if aggregates were formed, they were of low light scattering intensity with no apparent size evolution.

The weak interactions observed between AG and ST were in agreement with previous studies [3,6,8,10]. AG is composed of a ramified β(1 → 3)-D-galactose core that is highly branched at the 6 position with β(1 → 6) linked D-galactan side chains that are highly substituted with arabinose residues and to a lesser extent, glucuronic acid and rhamnose residues [29]. This highly branched structure and low mobility of galactan side chains may limit its ability to aggregate with tannin through hydrophobic interactions [6]. Application of commercial MP in red wine has been observed to either promote tannin aggregation and precipitation [37] or limit the loss of anthocyanin adducts [14]. Similarly, in model wine solution, a commercial MP (10% protein *w*/*w*, molecular weight distribution 14–500 kDa) has been observed to form large aggregates with grape and wine tannins [7], consistent with the current results. In contrast, MP purified from wine, in particular the low molecular weight fractions (1.6–3.5%, protein *w*/*w* with narrow molecular weight distribution around 51 to 62 kDa), limited seed tannin aggregation through steric hindrance, resulting in a smaller overall particle size [3,8]. The MP used in the current study had a molecular weight distribution between 10 and 98 kDa (Table 1) with a protein content at 11% of the dry weight. It appeared that the different behaviours towards ST were more related to protein content than molecular size. It has been shown that mannoproteins have significantly lower affinity to tannin than yeast-derived protein and bovine serum albumin [7,38]. Furthermore, between two wine AGP fractions, only the one with slightly higher protein content (3.6 vs. 0.8%) could form aggregates with procyanidins of DP 30 [6]. If such a small proportion of protein could induce a substantial difference in aggregate formation between polysaccharides and tannins, it might also explain the different behaviours between MP and AG in the current study, since MP had a higher protein content than AG (10 vs. 1.4%). This would potentially have significant implications for wine production. This is because native wine polysaccharide composition is highly variable and capable of impacting on tannin composition and subsequently wine astringency [13,39]. Furthermore commercial polysaccharide supplements could also be added to wine, as discussed previously, which adds further unknowns to the system. It has been shown that the protein content of commercial MP products can range from 10 to 50% [40]. Therefore, the choice of product, with varying concentration of active ingredients and other by-products, could have a great impact on the final wine colloidal state, potentially affecting the colour and organoleptic characters of wine post-addition.

Polysaccharide is thought to be important in mediating tannin and protein aggregation, through one or more mechanisms: (i) polysaccharides form ternary complexes with tannin-protein aggregates and thereby increase their solubility; and (ii) polysaccharides bind with tannin and thus limit access of protein [18,41]. The current study showed that for certain polysaccharides, the second mechanism is in effect. In the future, different types of protein could be introduced into this system to explore the competition between polysaccharides and various proteins for tannin binding in greater detail.

Lower ethanol concentration was found to promote aggregate formation between ST and MP. Ethanol preferentially solvates tannin and reducing ethanol content in an aqueous solution decreases tannin solubility [2,31]. Moreover, lowering ethanol concentration favors tannin-tannin or tannin-protein aggregation through hydrophobic interactions [31,42]. The effect of ethanol on tannin and mannoprotein interactions have been reported [8,38]. However, none of these studies reported an effect when the concentration differences between treatments were as small as those used in the current study (3%). Ethanol concentrations between 12 and 15% are typically found in red table wine. From a sensory point of view, a 4% increase in alcohol concentration could reduce astringency and enhance bitterness (two mouthfeel characters highly associated with wine polyphenolic composition) in model wine solutions [43,44]. Thus, the effect of ethanol on the colloidal state of wine macromolecules and its implication for wine sensory characters warrants further investigation.

3. Materials and Methods

3.1. Preparation of Polysaccharide and Tannin Materials

Cabernet Sauvignon grapes were harvested at the pre-veraison stage (green berries, pea size) from a commercial vineyard in South Australia, and frozen at −80 °C until used. Frozen berries were partially defrosted while kept on ice, and the seeds removed using a scalpel. A sample of 100 g of seeds was extracted overnight in 200 mL of 70% *v*/*v* aqueous acetone containing 10 mg/mL ascorbic acid. Extracts were filtered through a 0.5 mm mesh to remove solids and the recovered solution was centrifuged at 1730× *g*. Acetone was removed from the supernatant under vacuum at 35 °C and the remaining aqueous solution was lyophilized. The dried extract was reconstituted in 50 mL of 60% *v*/*v* HPLC grade aqueous methanol containing 0.05% *v*/*v* trifluoroacetic acid (TFA) and then applied (~18.3 mL/min) to a glass column (Michel-Miller, 300 × 21 mm, Vineland, NJ, USA) containing Sephadex LH20 chromatography resin (Amersham, Uppsala, Sweden) to an approximate bed volume of 93 mL, previously equilibrated with the loading solvent. The monomeric phenolics, organic acids and sugars were removed by application of 300 mL of 60% *v*/*v* aqueous methanol containing 0.05% *v*/*v* TFA. ST was recovered following application of 250 mL of 70% *v*/*v* aqueous acetone containing 0.05% *v*/*v* TFA. The eluted ST fraction was concentrated under reduced pressure at 35 °C to remove organic solvents and then lyophilized to a dry powder. ST was stored under nitrogen at −20 °C until used. The subunit composition of ST was determined by HPLC following acid catalysis in the presence of excess phloroglucinol [45,46]. The molar proportion of each subunit, mean degree of polymerization and mass conversion are reported in Supplementary Table S1.

Two polysaccharides were prepared from commercial supplements used in vinification. The MP product was a highly pure cell wall extract from *Saccharomyces cerevisiae* (Mannofeel, Laffort

Australia, Adelaide, Australia) while the arabinogalactan was purified from a commercial blend of gum arabic and grape tannin (Surli Vitis, Enartis Pacific, Melbourne, Australia), by removing the associated phenolic compounds with three extractions in 70% acetone (monitored by HPLC with a UV-vis detector at 280 nm). Both polysaccharides were dialyzed against 4 changes of MilliQ water using a 7 kDa cut-off membrane (SnakeSkin dialysis tubing, Thermo Scientific, Rockford, IL, USA), and then lyophilized. The subunit composition of polysaccharide was determined according to a published method [17]. Briefly, 1 mg/mL polysaccharide solution was hydrolysed in 2 M TFA for 3 h at 100 °C. Hydrolysates were dried in vacuo and reconstituted in 0.4 mL of Milli-Q water and mixed 1:1 with an aqueous internal standard solution comprising 0.6 mM ribose and deoxy-glucose (Sigma Aldrich, St. Louis, MO, USA). Mixtures were derivatized with 1-phenyl-3-methyl-5-pyrazolone (PMP) and analysed by RP-HPLC, using a C18 column (Kinetex, 2.6 µm, 100 Å, 100 × 3 mm). The HPLC instrumentation and mobile phase gradient were as reported previously (Bindon, et al. 2016). Total nitrogen content was measured by the analytical services unit of the Commonwealth Scientific and Industrial Research Organization (CSIRO, Adelaide, Australia), using a TruMAC (Leco Corporation, Saint Joseph, MI, USA); powdered polysaccharides were combusted in an atmosphere of oxygen and nitrogen determined as gaseous N2 by thermal conductivity detection. The composition of the products are reported in Supplementary Table S5.

Two model wine solutions (4 mg/mL tartaric acid, pH adjusted to 3.4 with analytical grade sodium hydroxide) containing ethanol levels at 12 and 15% (*v*/*v*) were used in the current study. The ionic strength was estimated at 20 mM based on the relative abundance of tartaric acid and its ionized forms at pH 3.4. Solutions were filtered through a 0.2 µm membrane (Durapore, Merck Millipore, Cork, Ireland) before use. For all experiments, ST, MP and AG were reconstituted in model wine solution at gravimetric concentrations (*w*/*v*).

3.2. Macromolecule Characterization

3.2.1. Size Exclusion Chromatography Analysis

The molecular weight distribution of ST, MP and AG were determined by size exclusion chromatography (SEC). ST was analysed with an HPLC (Agilent 1100, Agilent Technologies Australia Pty. Ltd., Melbourne, Australia), using the gel permeation chromatography (GPC) method originally reported by Kennedy and Taylor [46], with modifications described by Bindon and Kennedy [47]. The retention times at 10% and 90% ST elution by volume were compared against the standard curve to derive lower and upper ranges for molecular weight, while the retention time at 50% elution was used to determine mean molecular weight. In addition, the polydispersity index (PdI) was calculated by dividing weight average molecular weight (M_r) by number average molecular weight (M_n).

The size distribution of polysaccharides was analysed using an Agilent 1260 HPLC system fitted with a Yarra SEC-4000 column connected to a Yarra SEC-2000 column (silica resin, 3 µm, 300 × 7.8 mm, Phenomenex, Torrance, CA, USA). The mobile phase was 0.1 M NaNO3 with a flow rate of 1.2 mL/min for a 22.5 min run time, at 40 °C. Refractive index signals were analysed with ChemStation GPC data analysis software Rev B.01.01 (Agilent Technologies Australia Pty. Ltd., Melbourne, Australia). Polysaccharide molecular weight was determined by comparing samples to a calibration curve developed with a series of pullulan standards of known molecular weight (Shodex, Showa Denko K.k, Japan): P800 (708 kDa), P400 (344 kDa), P200 (200 kDa), P100 (107 kDa), P50 (47.1 kDa), P20 (21.1 kDa), P10 (7.6 kDa) and P5 (5.9 kDa). Each standard was run 5 times to check for retention time shift, which was not observed (data not shown). A third order polynomial curve was established between elution volume and molecular weight, with an R^2 of 0.9973 (Supplementary Figure S2). The mean and range of molecular weight, as well as PdI of polysaccharides, were determined in the same way as described for ST.

3.2.2. DLS Analysis

A Malvern Zetasizer Nano ZS (Malvern Instruments Ltd, Worcestershire, UK), equipped with a 4 mW He-Ne laser at a wavelength of 633 nm was used in the current study. Instrument control and data analysis were performed with Zetasizer software (version 7.10). For each measurement, the temperature was maintained at 25 °C, and the angle of detection was set at 90°. Measurement position, attenuator level and measurement duration were all set to be automatically optimized by the software.

Particle size (hydrodynamic diameter) was determined using the Stokes-Einstein equation:

$$d(H) = \frac{kT}{3\pi\eta D} \tag{1}$$

where k is Bolzmann's constant; T is absolute temperature; η is dispersant viscosity and D is diffusion coefficient. Viscosity was determined with Zetasizer software based on the molar content of ethanol in solutions and were 1.367 cP and 1.518 cP respectively for 12 and 15% ethanol model wines. D was determined by fitting an autocorrelation function to the exponential with two different algorithms: (i) cumulants analysis, which determined the mean particle size (Z-ave) and polydispersity index (PdI); and (ii) non-negative least squares (NNSL) analysis, which generated intensity weighted size distribution, using the 'general purpose mode' in this instance. Disposable low volume cuvettes with a pathlength of 10 mm were used for measurements.

3.2.3. NTA Analysis

Nanoparticle tracking analysis (NTA) was performed on a Nanosight NS300 (Malvern Instruments Ltd, Worcestershire, UK), equipped with a 635 nm laser and a scientific CMOS camera. NTA 3.0 software was used for instrument control and data analysis. The data was collected in the form of 60-second videos captured by the camera. The sample chamber was maintained at 25 °C and a syringe pump was used to keep a continuous flow of sample through the flow cell at 7 μL/min for the duration of measurement.

For each individual sample, settings (screen gain, camera level and focus) were manually adjusted to optimize visualization of the particles and thereafter kept identical for all video repetitions of the same sample (Supplementary Table S3). Detection threshold, which determined the minimal brightness of pixels to be considered for tracking, was also adjusted post-acquisition to minimize noise as well as to maintain a particle per frame count appropriate for analysis (10–100 particles per frame). Settings were kept consistent for all video repetitions of the same sample. The NTA software measured the mean square displacement from the centre of the particle's scatter as it moved from frame to frame in the collected videos. The hydrodynamic diameter of particles were calculated from the modified Einstein-Stokes equation:

$$\overline{(x,\ y)^2} = \frac{4kTt}{3\pi d\eta} \tag{2}$$

where $\overline{(x, y)^2}$ is the mean square of displacement; k is Bolzmann's constant; T is absolute temperature; t is time; d is the hydrodynamic diameter and η is dispersant viscosity [27]. The viscosity values were identical to those used in DLS measurements.

3.2.4. System Qualification for NTA and DLS Instruments

NIST-traceable polystyrene latex bead standards (100, 200 and 400 nm) were supplied by Malvern Instruments Ltd (Worcestershire, UK). The standards were dispersed in 0.01 M KCl. For DLS measurements, all three bead standards were diluted 1:10. For NTA measurements the dilution factors were according to the instrument supplier's manual, i.e., 1:1000 dilution for 100 nm, 1:100 dilution for 200 nm and 1:10 dilution for 400 nm. All samples were measured 5 times, by either DLS or NTA. For each system, the accuracy of measurements of 100 nm and 200 nm beads were within those specified by the International Standardization Organization [27,48] and were in good agreement with one another (Supplementary Table S6). Although the measurements for 400 nm beads deviated more from the stated size, accuracy was still within 10% for each method.

3.2.5. Particle Size of Tannin and Polysaccharide Determined by DLS and NTA

For particle size distribution determined by either DLS or NTA, ST, MP and AG were reconstituted by vortexing in model wine containing 12% ethanol. All solutions were centrifuged at 3273× *g* for 5 min before measurements. Polysaccharides and ST were reconstituted in model wine at 0.5 mg/mL and 0.125 mg/mL respectively, for NTA characterization. At these concentrations no excessive scattering was observed while all particles could be clearly visualized under the scientific CMOS camera. Fifteen video repetitions were taken for each sample. DLS analysis required samples to be much more concentrated. Higher concentrations were trialled on DLS to find a working concentration that was closest to those used for NTA analysis. It was found that 4 mg/mL was the minimal concentration at which sufficient scattered light could be detected by the DLS instrument during a measurement, i.e., a mean count rate higher than 20 kilo counts per second and therefore this concentration was chosen. Each sample was measured five times.

3.3. *Characterization of Interactions between Polysaccharides and Tannins*

3.3.1. UV-visible Spectroscopy Analysis

The aggregation between polysaccharides and tannins at various concentrations were measured as absorbance at 650 nm by UV-visible spectrometry. This assay was adapted and modified from a previous study [6]. ST was reconstituted in the two model wine solutions at 10 mg/mL, while MP and AG were reconstituted separately at 1 mg/mL. The control samples consisted of 1 mL of diluted ST solution of 0, 0.078, 0.156, 0.313, 0.625, 1.25, 2.5 and 5 mg/mL (*w/v*), along the columns on a 96-well plate (1.1 mL volume, Axygen, Adelab, Adelaide, Australia). For the treatment samples, 0.5 mg/mL of either MP or AG was added to the ST solutions, while maintaining the same tannin concentrations and volumes as control samples. Both control and treatment samples were prepared in duplicate. The plates were sealed with a compatible silicone sealing mat, vigorously shaken and stored at 22 °C for 24 h. Thereafter, 200 µL from each well was then transferred into a clear 96-well cycloolefine plate (Greiner, Sigma-Aldrich, Sydney, Australia) and the absorbance at 650 nm recorded with a SpectraMax M2 Microplate reader (Molecular Devices, Melbourne, Australia). In addition, a 20 µL sample aliquot was diluted with 980 µL of 1 M HCl solution and 280 nm absorbance was recorded to determine total phenolics, according to Mercurio, Dambergs, Herderich, and Smith [49]. The plate was then centrifuged at 3273× *g* for 5 min, and another 20 µL sample diluted with 980 µL of 1 M HCl and the absorbance measured at 280 nm.

3.3.2. NTA and DLS Analyses

ST was reconstituted at 2.5 mg/mL or 10 mg/mL, while AG and MP were both reconstituted at 1 mg/mL, in each model wine solution. The ST solution was mixed in equal parts (750 µL each) with each polysaccharide solution and all stock solutions were diluted 1:1 with model wine to create two series of samples with final concentrations of (i) 1.25 mg/mL tannin, 0.5 mg/mL polysaccharide and their mixtures and (ii) 5 mg/mL tannin, 0.5 mg/mL polysaccharide and their mixtures. The solutions were sealed in 1.5 mL Eppendorf tubes and kept at 22 °C for 24 h before being centrifuged at 16,100× *g* for 5 min to remove insoluble particles, if any, as shown by others [7,32]. The samples were then used directly for DLS analysis. However, for NTA, the supernatants containing ST, individually or combined with either polysaccharide type, were diluted 1 in 10 with model wine solutions for the low concentration series and 1 in 40 for the high concentration series, while supernatants containing only polysaccharide were measured without dilution. The samples for NTA and DLS measurements were individually prepared. For all samples, 15 video repetitions were recorded by NTA and four replicates were performed by DLS.

3.3.3. Tannin Solubility

It was considered that large tannin aggregates might be present due to insolubility, according to previous studies [31,32]. The Cabernet Sauvignon seed tannin isolate used for the experiments (mass conversion 67% *w/w* by phloroglucinolysis) was prepared in 15% acidified ethanol (model wine) as described above, or made up directly in pure ethanol, then diluted to 15% *v/v* with water containing tartaric acid, equivalent to the model wine matrix used for experimentation. NTA of the samples was then compared when analysed neat, or centrifuged at 16,100 g for 10 min which according to Zanchi and colleagues [32] would be adequate to sediment insoluble, unstable tannin particles. For tannin solutions directly reconstituted in model wine, no difference in particle concentration was observed following centrifugation, but there was a small increase in particle size. For tannin reconstituted in ethanol and then diluted, higher particle concentrations were observed than for direct reconstitution which were lost (50%) upon centrifugation (data not shown), indicating this mode of reconstitution produced colloidal instability. Hence, samples were reconstituted directly in model wine for all experiments.

4. Conclusions

This study presents the first application of NTA for the characterization of tannin and polysaccharide interactions in wine-like media. NTA was able to assess shifts in size distribution in aggregated colloids, following addition of commercial polysaccharide supplements into model wine solutions containing grape seed tannins. The light scattering intensity of individually tracked particles can provide additional insight into aggregate formation. The NTA results were confirmed by DLS and UV-vis analysis. The two polysaccharides, MP and AG, derived from commercial winemaking additives used in wine production, were considerably different in colloidal behaviour when mixed with ST. MP formed larger, highly light scattering aggregates, while AG had only weak interactions with ST, with no clear indications of aggregate formation. A 3% ethanol reduction was found to increase aggregate size for MP, but had no impact on AG.

Supplementary Materials: The following are available online at http://www.mdpi.com/1420-3049/24/11/2100/s1. Table S1. Subunit composition of ST; Table S2. Absorbance (280 nm) of ST at different concentrations, either individually or combined with 0.5 mg/mL polysaccharide (MP or AG), in 12% and 15% ethanol model wine, before and after centrifugation; Table S3. Camera shutter and gain settings for binding experiment characterized by NTA; Table S4. Polydispersity index (PdI) and intensity weighted mean particle size distribution determined by dynamic light scattering. The samples contained ST at either 1.25 or 5 mg/mL, combined with 0.5 mg/mL of either MP or AG; Table S5. Monosaccharide residue composition of polysaccharide following hydrolysis; Table S6. Mean size and size distribution of polystyrene beads determined by dynamic light scattering and nanoparticle tracking analysis; Figure S1. Comparison of light scattering intensity (measured as derived count of photons) of ST combined with either MP or AGP, in both 12% and 15% model wine solutions; Figure S2. Calibration curve for polysaccharide molecular weight based on size exclusion chromatography.

Author Contributions: Conceptualization, S.L. and K.A.B.; methodology, S.L., A.M.-V. and K.A.B.; data analysis, S.L. and K.A.B.; resources, K.L.W.; writing—original draft preparation, S.L. and K.A.B.; writing—review and editing, S.L., K.L.W., A.M.-V. and K.A.B.; supervision, K.L.W. and K.A.B.; project administration, K.L.W. and K.A.B.; funding acquisition, K.L.W. and K.A.B.

Funding: This research was conducted within the Waite Precinct by the Australian Research Council Industrial Transformation Training Centre for Innovative Wine Production (project number IC130100005) and the Australian Wine Research Institute (AWRI). The AWRI, a member of the Wine Innovation Cluster in Adelaide, is supported by Australian grapegrowers and winemakers through their investment body, Wine Australia, with matching funds from the Australian Government. Sijing Li is also the recipient of a Wine Australia supplementary scholarship (GWR Ph1312).

Acknowledgments: We thank Laffort Australia and Enartis Pacific for their generous supply of materials for this study. From the AWRI, Tim Reilly is thanked for support in performing NTA analysis of seed tannin samples, and Josh Hixson is thanked for purification of the Tannat seed tannin extract.

Conflicts of Interest: The authors declare no conflict of interest.

References

1. Smith, P.A.; McRae, J.M.; Bindon, K.A. Impact of winemaking practices on the concentration and composition of tannins in red wine. *Aust. J. Grape Wine Res.* **2015**, *21*, 601–614. [CrossRef]
2. Poncet-Legrand, C.; Cartalade, D.; Putaux, J.L.; Cheynier, V.; Vernhet, A. Flavan-3-ol aggregation in model ethanolic solutions: Incidence of polyphenol structure, concentration, ethanol content, and ionic strength. *Langmuir* **2003**, *19*, 10563–10572. [CrossRef]
3. Riou, V.; Vernhet, A.; Doco, T.; Moutounet, M. Aggregation of grape seed tannins in model wine—Effect of wine polysaccharides. *Food Hydrocoll.* **2002**, *16*, 17–23. [CrossRef]
4. McRae, J.M.; Mierczynska-Vasilev, A.; Soden, A.; Barker, A.M.; Day, M.P.; Smith, P.A. Effect of commercial-scale filtration on sensory and colloidal properties of red wines over 18 months bottle aging. *Am. J. Enol. Vitic.* **2017**, *68*, 263–274. [CrossRef]
5. Waters, E.J.; Peng, Z.; Pocock, K.F.; Jones, G.P.; Clarke, P.; Williams, P.J. Solid-State ^{13}C-NMR investigation into insoluble deposits adhering to the inner glass surface of bottled red wine. *J. Agric. Food Chem.* **1994**, *42*, 1761–1766. [CrossRef]
6. Watrelot, A.A.; Le Bourvellec, C.; Imberty, A.; Renard, C.M.G.C. Neutral sugar side chains of pectins limit interactions with procyanidins. *Carbohydr. Polym.* **2014**, *99*, 527–536. [CrossRef]
7. Mekoue Nguela, J.; Poncet-Legrand, C.; Sieczkowski, N.; Vernhet, A. Interactions of grape tannins and wine polyphenols with a yeast protein extract, mannoproteins and β-glucan. *Food Chem.* **2016**, *210*, 671–682. [CrossRef]
8. Poncet-Legrand, C.; Doco, T.; Williams, P.; Vernhet, A. Inhibition of grape seed tannin aggregation by wine mannoproteins: Effect of polysaccharide molecular weight. *Am. J. Enol. Vitic.* **2007**, *58*, 87–91.
9. Guadalupe, Z.; Ayestarán, B.; Williams, P.; Doco, T. Determination of must and wine polysaccharides by gas chromatography-mass spectrometry (GC-MS) and size-exclusion chromatography (SEC). In *Polysaccharides: Bioactivity and Biotechnology*; Ramawat, K., Mérillon, J.M., Eds.; Springer International Publishing: Basel, Switzerland, 2015; pp. 1265–1297.
10. Carvalho, E.; Mateus, N.; Plet, B.; Pianet, I.; Dufourc, E.; De Freitas, V. Influence of wine pectic polysaccharides on the interactions between condensed tannins and salivary proteins. *J. Agric. Food Chem.* **2006**, *54*, 8936–8944. [CrossRef]
11. Mateus, N.; Carvalho, E.; Luís, C.; De Freitas, V. Influence of the tannin structure on the disruption effect of carbohydrates on protein-tannin aggregates. *Anal. Chim. Acta* **2004**, *513*, 135–140. [CrossRef]
12. Rinaldi, A.; Gambuti, A.; Moio, L. Precipitation of salivary proteins after the interaction with wine: the effect of ethanol, pH, fructose, and mannoproteins. *J. Food Sci.* **2012**, *77*, 485–490. [CrossRef] [PubMed]
13. Watrelot, A.A.; Schulz, D.L.; Kennedy, J.A. Wine polysaccharides influence tannin-protein interactions. *Food Hydrocoll.* **2017**, *63*, 571–579. [CrossRef]
14. Alcalde-Eon, C.; García-Estévez, I.; Puente, V.; Rivas-Gonzalo, J.C.; Escribano-Bailón, M.T. Color stabilization of red wines. A chemical and colloidal approach. *J. Agric. Food Chem.* **2014**, *62*, 6984–6994. [CrossRef] [PubMed]
15. Maury, C.; Sarni-Manchado, P.; Poinsaut, P.; Cheynier, V.; Moutounet, M. Influence of polysaccharides and glycerol on proanthocyanidin precipitation by protein fining agents. *Food Hydrocoll.* **2016**, *60*, 598–605. [CrossRef]
16. Australian Government Standard 4.5.1—Wine Production Requirements (Australia Only), Version F2017C01001. Australia New Zealand Food Standards Code: 2014. Available online: https://www.legislation.gov.au/Details/F2017C01001 (accessed on 24 April 2019).
17. Bindon, K.A.; Carew, A.L.; Mierczynska-Vasilev, A.; Kassara, S.; Kerslake, F.; Smith, P.A. Characterization of macromolecular complexes in red wine: Composition, molecular mass distribution and particle size. *Food Chem.* **2016**, *199*, 838–846. [CrossRef] [PubMed]
18. Brandão, E.; Silva, M.S.; García-Estévez, I.; Williams, P.; Mateus, N.; Doco, T.; de Freitas, V.; Soares, S. The role of wine polysaccharides on salivary protein-tannin interaction: A molecular approach. *Carbohydr. Polym.* **2017**, *177*, 77–85. [CrossRef]

19. Carn, F.; Guyot, S.; Baron, A.; Pérez, J.; Buhler, E.; Zanchi, D. Structural properties of colloidal complexes between condensed tannins and polysaccharide hyaluronan. *Biomacromolecules* **2012**, *13*, 751–759. [CrossRef] [PubMed]
20. Mamet, T.; Ge, Z.Z.; Zhang, Y.; Li, C.M. Interactions between highly galloylated persimmon tannins and pectins. *Int. J. Biol. Macromol.* **2018**, *106*, 410–417. [CrossRef] [PubMed]
21. Soares, S.I.; Gonçalves, R.M.; Fernandes, I.; Mateus, N.; de Freitas, V. Mechanistic approach by which polysaccharides inhibit α-amylase/procyanidin aggregation. *J. Agric. Food Chem.* **2009**, *57*, 4352–4358. [CrossRef]
22. Watrelot, A.A.; Le Bourvellec, C.; Imberty, A.; Renard, C.M.G.C. Interactions between pectic compounds and procyanidins are influenced by methylation degree and chain length. *Biomacromolecules* **2013**, *14*, 709–718. [CrossRef] [PubMed]
23. Filipe, V.; Hawe, A.; Jiskoot, W. Critical evaluation of nanoparticle tracking analysis (NTA) by NanoSight for the measurement of nanoparticles and protein aggregates. *Pharm. Res.* **2010**, *27*, 796–810. [CrossRef]
24. Jarzębski, M.; Bellich, B.; Białopiotrowicz, T.; Śliwa, T.; Kościński, J.; Cesàro, A. Particle tracking analysis in food and hydrocolloids investigations. *Food Hydrocoll.* **2017**, *68*, 90–101. [CrossRef]
25. Bindon, K.A.; Smith, P.A.; Holt, H.; Kennedy, J.A. Interaction between grape-derived proanthocyanidins and cell wall material. 2. implications for vinification. *J. Agric. Food Chem.* **2010**, *58*, 10736–10746. [CrossRef]
26. Technical Note: Size Quality Report for the Zetasizer Nano. Available online: http://www.malvern.com/en/support/resource-center/technical-notes/TN101104SizeQualityReportZetasizerNano.aspx (accessed on 3 April 2017).
27. ISO 19430:2016 Particle Size Analysis—Particle Tracking Analysis (PTA) Method. 2016. Available online: https://www.iso.org/standard/64890.html (accessed on 24 April 2019).
28. Al Assaf, S.; Phillips, G.O.; Williams, P.A. Studies on acacia exudate gums. Part I: the molecular weight of Acacia senegal gum exudate. *Food Hydrocoll.* **2005**, *19*, 647–660. [CrossRef]
29. Li, X.; Fang, Y.; Al-Assaf, S.; Phillips, G.O.; Nishinari, K.; Zhang, H. Rheological study of gum arabic solutions: Interpretation based on molecular self-association. *Food Hydrocoll.* **2009**, *23*, 2394–2402. [CrossRef]
30. Lankhorst, P.P.; Voogt, B.; Tuinier, R.; Lefol, B.; Pellerin, P.; Virone, C. Prevention of tartrate crystallization in wine by hydrocolloids: The mechanism studied by Dynamic light scattering. *J. Agric. Food Chem.* **2017**, *65*, 8923–8929. [CrossRef] [PubMed]
31. Zanchi, D.; Vernhet, A.; Poncet-Legrand, C.; Cartalade, D.; Tribet, C.; Schweins, R.; Cabane, B. Colloidal Dispersions of tannins in water−ethanol solutions. *Langmuir* **2007**, *23*, 9949–9959. [CrossRef] [PubMed]
32. Zanchi, D.; Konarev, P.V.; Tribet, C.; Baron, A.; Svergun, D.I.; Guyot, S. Rigidity, conformation, and solvation of native and oxidized tannin macromolecules in water-ethanol solution. *J. Chem. Phys.* **2009**, *130*, 245103. [CrossRef]
33. Poncet-Legrand, C.; Cabane, B.; Bautista-Ortín, A.-B.; Carrillo, S.; Fulcrand, H.; Pérez, J.; Vernhet, A. Tannin oxidation: Intra- versus intermolecular reactions. *Biomacromolecules* **2010**, *11*, 2376–2386. [CrossRef] [PubMed]
34. Cheah, K.Y.; Howarth, G.S.; Bindon, K.A.; Kennedy, J.A.; Bastian, S.E.P. Low molecular weight procyanidins from grape seeds enhance the impact of 5-Fluorouracil chemotherapy on Caco-2 human colon cancer cells. *PLoS ONE* **2014**. [CrossRef] [PubMed]
35. Hixson, J.L.; Bindon, K.A.; Smith, P.A. Evaluation of Direct Phloroglucinolysis and colorimetric depolymerization assays and their applicability for determining condensed tannins in grape marc. *J. Agric. Food Chem.* **2015**, *63*, 9954–9962. [CrossRef] [PubMed]
36. Le Bourvellec, C.; Renard, C.M.G.C. Interactions between polyphenols and macromolecules: quantification methods and mechanisms. *Crit. Rev. Food Sci. Nutr.* **2012**, *52*, 213–248. [CrossRef] [PubMed]
37. Guadalupe, Z.; Ayestarán, B. Effect of commercial mannoprotein addition on polysaccharide, polyphenolic, and color composition in red wines. *J. Agric. Food Chem.* **2008**, *56*, 9022–9029. [CrossRef]
38. Rowe, J.D.; Harbertson, J.F.; Osborne, J.P.; Freitag, M.; Lim, J.; Bakalinsky, A.T. Systematic identification of yeast proteins extracted into model wine during aging on the yeast lees. *J. Agric. Food Chem.* **2010**, *58*, 2337–2346. [CrossRef] [PubMed]
39. Quijada-Morín, N.; Williams, P.; Rivas-Gonzalo, J.C.; Doco, T.; Escribano-Bailón, M.T. Polyphenolic, polysaccharide and oligosaccharide composition of Tempranillo red wines and their relationship with the perceived astringency. *Food Chem.* **2014**, *154*, 44–51. [CrossRef] [PubMed]

40. Li, S.; Wilkinson, K.L.; Bindon, K.A. Compositional variability in commercial tannin and mannoprotein products. *Am. J. Enol. Vitic.* **2018**, *69*, 176–181. [CrossRef]
41. Scollary, G.R.; Pásti, G.; Kállay, M.; Blackman, J.; Clark, A.C. Astringency response of red wines: Potential role of molecular assembly. *Trends Food Sci. Technol.* **2012**, *27*, 25–36. [CrossRef]
42. McRae, J.M.; Ziora, Z.M.; Kassara, S.; Cooper, M.A.; Smith, P.A. Ethanol concentration influences the mechanisms of wine tannin interactions with poly(l-proline) in model wine. *J. Agric. Food Chem.* **2015**, *63*, 4345–4352. [CrossRef]
43. Fontoin, H.; Saucier, C.; Teissedre, P.L.; Glories, Y. Effect of pH, ethanol and acidity on astringency and bitterness of grape seed tannin oligomers in model wine solution. *Food Qual. Preference* **2008**, *19*, 286–291. [CrossRef]
44. Vidal, S.; Courcoux, P.; Francis, L.; Kwiatkowski, M.; Gawel, R.; Williams, P.; Waters, E.; Cheynier, V. Use of an experimental design approach for evaluation of key wine components on mouth-feel perception. *Food Qual. Preference* **2004**, *15*, 209–217. [CrossRef]
45. Kennedy, J.A.; Jones, G.P. Analysis of proanthocyanidin cleavage products following acid-catalysis in the presence of excess phloroglucinol. *J. Agric. Food Chem.* **2001**, *49*, 1740–1746. [CrossRef] [PubMed]
46. Kennedy, J.A.; Taylor, A.W. Analysis of proanthocyanidins by high-performance gel permeation chromatography. *J. Chromatogr. A* **2003**, *995*, 99–107. [CrossRef]
47. Bindon, K.A.; Kennedy, J.A. Ripening-induced changes in grape skin proanthocyanidins modify their interaction with cell walls. *J. Agric. Food Chem.* **2011**, *59*, 2696–2707. [CrossRef]
48. ISO 22412:2008 Particle size analysis—Dynamic light scattering (DLS). *International Organization of Standardization*. 2008. Available online: https://www.iso.org/standard/40942.html (accessed on 24 April 2019).
49. Mercurio, M.D.; Dambergs, R.G.; Herderich, M.J.; Smith, P.A. High throughput analysis of red wine and grape phenolics—Adaptation and validation of methyl cellulose precipitable tannin assay and modified Somers color assay to a rapid 96 well plate format. *J. Agric. Food Chem.* **2007**, *55*, 4651–4657. [CrossRef] [PubMed]

Sample Availability: Sample of the purified seed tannin is available from the authors. Commercial mannoprotein and gum arabic products are available from the producing companies (details are provided in the Materials and Methods section).

molecules

MDPI

Article

Madeira Wine Volatile Profile. A Platform to Establish Madeira Wine Aroma Descriptors

Rosa Perestrelo [1], Catarina Silva [1] and José S. Câmara [1,2,*]

1 CQM—Centro de Química da Madeira, Universidade da Madeira, Campus da Penteada, 9020-105 Funchal, Portugal
2 Departamento de Química, Faculdade de Ciências Exatas e Engenharia, Universidade da Madeira, Campus da Penteada, 9020-105 Funchal, Portugal
* Correspondence: jsc@staff.uma.pt; Tel.: +351-291705112

Academic Editors: Rosa Maria de Sá Perestrelo and José Sousa Câmara
Received: 5 July 2019; Accepted: 17 August 2019; Published: 21 August 2019

Abstract: In the present study we aimed to investigate the volatile organic compounds (VOCs) that may potentially be responsible for specific descriptors of Madeira wine providing details about Madeira wine aroma notes at molecular level. Moreover, the wine aroma profile, based on the obtained data, will be a starting point to evaluate the impact of grape variety (Malvasia, Bual, Sercial, Verdelho and Tinta Negra), type (sweet, medium sweet, dry and medium dry), and age (from 3 to 20 years old) on Madeira wine sensorial properties. Firstly, a comprehensive and in-depth Madeira wine volatile profiling was carried out using headspace solid-phase microextraction combined with gas chromatography-mass spectrometry (HS–SPME/GC–qMS). Secondly, a relation among the varietal, fermentative and aging aroma compounds, and their aroma descriptors with the Madeira wine sensorial properties was assessed. A total of 82 VOCs, belonging to different chemical families were identified, namely 21 esters, 13 higher alcohols, ten terpenic compounds, nine fatty acids, seven furanic compounds, seven norisoprenoids, six lactones, four acetals, four volatile phenols and one sulphur compound. From a sensorial point of view, during the aging process the wine lost its freshness and fruitiness odor related to the presence of some varietal and fermentative compounds, whereas other descriptors such as caramel, dried fruits, spicy, toasty and woody, arose during ageing. The Maillard reaction and diffusion from the oak were the most important pathways related with these descriptors. A relationship-based approach was used to explore the impact of grape variety, wine type, and age on Madeira wine sensorial properties based on shared number of VOCs and their odors.

Keywords: wine; VOCs; potential odorants; HS–SPME; GC–qMS

1. Introduction

Madeira wine is a fortified Portuguese wine produced in Madeira Island over the last centuries playing an important role in the Island economy. The specific characteristics of Madeira wine arise from a set of specific conditions including the terroir, unique grape varieties and the singular winemaking process. The fermentation process is stopped by the addition of natural grape spirit in order to obtain an ethanol content of 18–22% (*v*/*v*). Some wines are submitted to an aging process in oak casks, in cellars, at temperatures up to 30 °C, and humidity levels between 70 and 75%, while the majority of wines are submitted to a baking process, i.e., the wine is placed in large coated vats and the temperature is slowly increased at about 5 °C per day, and maintained at 45–50 °C for at least 3 months. After this treatment, the wine is allowed to undergo a maturation process in oak casks for a minimum of 3 years [1,2].

Wine volatile composition plays an important role in wine quality since it promotes several sensations during wine consumption, odors (due to molecules that can bind olfactory receptors) and can affect flavor (combination of aroma and taste) in mouth retro-nasally, that lead to consumer acceptance or rejection. The wine aroma consists of a combination of several hundred of different volatile organic compounds (VOCs), most of which are present in trace amounts (usually at μg/L or ng/L level) [3]. Nevertheless, the presence of a molecule at a concentration above its odor threshold (OT), is sufficient to provide a characteristic product aroma (impact odorant). Nevertheless, even when present at concentrations below their OTs, may contribute to the overall wine aroma, as a result of the interactions with other molecules [4].

Different extraction techniques, such as solid phase extraction (SPE) [5–9] and liquid-liquid extraction [5,10–13] have been applied on the establishment of volatile profile of Madeira wine. However, most of these approaches present several disadvantages, such as time- and labor-intensive, large solvent and sample amount, which can lead to analyze losses and a reduction in sensitivity. Currently, the trend in the analyze of VOCs is more focused in the use of miniaturized sample preparation, increasing of efficiency of analysis, no solvent techniques, such as solid-phase microextraction (SPME) [7,10,14,15], stir bar sorptive extraction (SBSE) [14] and microextraction by packed sorbents (MEPS) [7] followed by gas chromatography-mass spectrometry (GC-MS) process have been used for that purpose. GC-MS is user-friendly, fast, selective and very sensitive method to establish the volatile signature of several food matrices. In addition, it was also equipped with powerful data systems that are used not only to control and acquire data from the GC and MS, but also to identify flavor components by automated matching against reference libraries of spectra of known odorants [16].

The Madeira wines volatile composition has been topic of several studies, as the data obtained has been useful in the elucidation of basic flavor chemistry. Enormous efforts were focused on the topic of varietal (e.g., terpenoids, norisoprenoids) [14,17,18], pre-fermentative (e.g., C_6 alcohols and aldehydes) [11], fermentative (e.g., alcohols, esters, acids, carbonyl compounds) [5,8,14], and finally aging aroma compounds (e.g., volatiles extracted from oak, like volatile phenols, lactones) [5,8,15,19]. Few studies have focused on the establishment of potential impact odorants, which could contribute individually to the Madeira wine aroma [6]. Campo et al. [6] build a hierarchical list of the odorants using gas chromatography-olfactometry (GC-O) that express the aroma of Madeira wines. The GC–O profile of Madeira wines lacks on varietal compounds (e.g., terpenoids, cystein-derived thiols), is rich in sotolon, phenylacetaldehyde, (*Z*)-whiskey lactone and of some volatile phenols (e.g., guaiacol, 4-vinylguaiacol, *m*-cresol). Madeira wines contain a huge number of intense odorants not identified which were not even detected in the corresponding young wines [6]. Sotolon has also been previously reported as powerful odorant, responsible for burnt, curry, honey, nutty, spicy, walnut odors of Madeira wines, depending on their concentration and enantiomeric distribution [12,20].

This research aimed to provide details about Madeira wine aroma notes at molecular level, as it can be useful to explain its peculiar aroma. Moreover, the wine aroma profile is a natural starting point for a systematic search for principles to evaluate the impact of grape variety, type, and age on Madeira wine sensorial properties. Therefore, Madeira wines from different varieties (Malvasia, Bual, Sercial, Verdelho and Tinta Negra), types (sweet, medium sweet, dry and medium dry), and ages (from 3 to 20 years old) were analyzed. Firstly, in-depth Madeira wine volatile profiling (e.g., terpenic compounds, norisoprenoids, sulphur compound, alcohols, esters, lactones, furanic compounds, acetals) was carried out using HS–SPME/GC–qMS. Secondly, a relation among the varietal, fermentative and aging aroma compounds, and their aroma descriptors with the Madeira wine sensorial properties was performed. A relationship-based approach was used to explore the impact of grape variety, wine type, and age on Madeira wine sensorial properties based on shared number of VOCs and their odor descriptors.

2. Results and Discussion

HS–SPME/GC–qMS methodology was used to establish the Madeira wine volatile profiling, as a sensitive technique to explain the unique aroma descriptors of Madeira wines. Considering the five grape varieties under study, a set of 82 VOCs (Table 1), namely 21 esters, 13 higher alcohols, ten terpenic compounds, nine fatty acids, seven norisoprenoids, seven furanic compounds, six lactones, four acetals, four volatile phenols and one sulphur compound (Tables S1 and S2), have been identified by matching the obtained mass spectra with the reference compounds spectra in NIST Mass Spectral Search Program with a resemblance percentage above 80% and by comparison of the KIs calculated (KI_{calc}) with the values reported in the literature (KI_{lit}) for polyethylene glycol (or equivalent) column.

A range between 0 and 35 ($|KI_{calc} - KI_{lit}|$) was obtained for KI_{cal} compared to the KI_{lit} reported in the literature for GC with polyethylene glycol GC column or equivalent. This difference in KI is acceptable (<5%) taking into account that the literature data is obtained from a large range of GC stationary phases (several commercial GC columns are composed of polyethylene glycol or equivalent stationary phases). The relative concentration of each VOC and their relative standard deviation (%RSD) in dry/medium dry and sweet/medium sweet are available as Supplementary Material on Tables S1 and S2, respectively. Sixty-nine VOCs were common in all wine samples analyzed, namely seven terpenic compounds, six norisoprenoids, 13 alcohols, 21 esters, 9 acids, four acetals, seven furanic compounds and two volatile phenols (Tables S1 and S2).

The number of identified VOCs ranged from 77 to 79 for Malvasia wines, from 78 to 79 for Bual wines, from 76 to 80 for Sercial wines, and from 77 to 80 for Verdelho wines, for young and old wines, respectively. For Tinta Negra variety, different types of wines were considered. It was observed that the number of VOCs ranged from 77 to 79 for dry, from 80 to 82 for medium dry, from 76 to 80 for sweet, and from 79 to 81 for medium sweet, for young and old wines, respectively.

Table 1. Volatile organic compounds (VOCs) identified in Madeira wines by headspace solid-phase microextraction tandem with gas chromatography-mass spectrometry (HS–SPME/GC–qMS), and the corresponding odor thresholds and odor descriptors.

RT (min) [1]	KI_{calc} [2]	KI_{lit} [3]	ID [4]	Chemical Families	OTs (µg/L) [5]	Odor descriptors [6]
				Terpenic compounds		
7.34	1150	1158	MS, RI, Std	β-Pinene	6	[7] **Citrus**, **floral**, fruit, green, pine, sweet, terpenic, **wood**
8.38	1178	1182	MS, RI, Std	**Limonene** [8]	10	**Citrus**, fruit, **wood**
18.96	1430	1433	MS, RI	**(*E*)-Linalool oxide**	500	**Floral**, green, rose, sweet
19.16	1445	1451	MS, RI	**(*Z*)-Linalool oxide**	500	**Floral**, green, rose, sweet
22.88	1537	1537	MS, RI, Std	**Linalool**	15	**Citrus**, lavender, **floral**, fruit, green, muscat, sweet
24.01	1561	1566	MS, RI, Std	**β-Caryophyllene**	- [9]	Fruit, green, **spice**, **wood**
28.58	1673	1669	MS, RI, Std	**α-Terpineol**	250	Anise, **floral**, fruit, mint, oil, toothpaste
31.38	1764	1762	MS, RI, Std	Citronellol	30	**Citrus**, clove, **floral**, fresh, green, rose, sour, sweet
39.86	1981	2009	MS, RI, Std	**Geraniol**	20	**Citrus**, **floral**, fruit, **waxy**
41.99	2125	2134	MS, RI	δ-Cadinol	-	**Almond**, green, **waxy**, **wood**
				Norisoprenoids		
21.47	1498	1507	MS, RI, Std	**Vitispirane I**	800	Camphor, eucalyptus, **spice**, **wood**
21.57	1501	1510	MS, RI, Std	**Vitispirane II**	-	Camphor, eucalyptus, **spice**, **wood**
26.22	1614	1623	MS, RI, Std	**β-Cyclocitral**	5	**Floral**, sweet
30.86	1742	1755	MS, RI, Std	**TDN** [10]	2	**Floral**, fruit, pleasant, wine
32.33	1785	1790	MS, RI, Std	**β-Damascenone** [6]	0.05	**Floral**, fruit, **honey**, sweet, **tobacco**
34.79	1844	1840	MS, RI, Std	**Geranyl acetone**	60	**Floral**, fruit, green, **waxy**, **wood**
36.34	1910	1912	MS, RI, Std	β-Ionone	0.10	**Floral**, violet
				Higher Alcohols		
4.78	1074	1057	MS, RI, Std	**Butan-2-ol**	-	Alcohol, oil, wine
6.12	1113	1112	MS, RI, Std	**2-Methylpropan-1-ol**	40,000	Alcohol, bitter, glue, leek, licorice
7.87	1165	1176	MS, RI, Std	**Hexan-2-ol**	-	Fatty, fruit, wine
9.51	1206	1206	MS, RI, Std	**3-Methylbutan-1-ol**	30,000	Alcohol, balsamic, burnt, cheesy, fruit, pungent, ripe onion
15.15	1350	1354	MS, RI, Std	**Hexan-1-ol**	8000	**Floral**, fruit, green, herbal, mild, **toasty**, sweet, **wood**
15.23	1352	1362	MS, RI, Std	**(*E*)-3-Hexen-1-ol**	400	Fresh, green, grass, leaf
16.11	1371	1379	MS, RI, Std	**(*Z*)-3-Hexen-1-ol**	400	Fresh, green, grass, leaf
22.25	1514	1510	MS, RI, Std	**2-Ethylhexan-1-ol**	270	**Citrus**, fresh, **floral**, oil, sweet
22.43	1521	1524	MS, RI	**(*R*,*S*)-Butan 2,3-diol**	120,000	Fruit
23.93	1556	1556	MS, RI	**(*R*,*R*)-Butan-2,3-diol**	120,000	Fruit
32.29	1784	1783	MS, RI, Std	**Decan-1-ol**	400	Fatty
35.31	1861	1869	MS, RI, Std	**Benzyl alcohol**	200,000	Blackberry, **floral**, fruit
36.42	1915	1910	MS, RI, Std	**2-Phenyethyl alcohol**	14,000	**Floral**, herbal, **honey**, pollen, rose, **spice**, sweet
				Sulphur compound		
29.65	1701	1723	MS, RI, Std	Ethyl 3-(methylthio)propionate	7	Fruit, pineapple

Table 1. *Cont.*

RT (min) [1]	KI_{calc} [2]	KI_{lit} [3]	ID [4]	Chemical Families	OTs (μg/L) [5]	Odor descriptors [6]
				Esters		
4.68	1055	1047	MS, RI, Std	**Ethyl butanoate**	20	Acetone, bubblegum, **caramel**, fruit
4.71	1061	1053	MS, RI, Std	**Ethyl 3-methylbutanoate**	3	Anise, apple, blackcurrant, **citrus**, fruit, sweet
6.51	1125	1120	MS, RI, Std	**Isoamyl acetate**	30	**Banana**, fresh, fruit, sweet
10.04	1222	1220	MS, RI, Std	**Ethyl hexanoate**	5	Anise, **caramel**, fruit, wine
10.31	1254	1262	MS, RI, Std	**Hexyl acetate**	10	Acid, **citrus**, fruit, green, herbal, rubber, **spice**, **tobacco**
10.52	1279	1276	MS, RI, Std	**Ethyl pyruvate**	-	**Caramel**, **ethereal**, fruit, vegetable, sweet
14.67	1339	1340	MS, RI	**Ethyl lactate**	154,636	Acidic, **ethereal**, fruit, sweet
18.16	1416	1414	MS, RI, Std	**Ethyl octanoate**	2	Fruit, must, soap, sweet, **waxy**
20.98	1478	1483	MS, RI, Std	**Ethyl 3-hydroxybutanoate**	20,000	Coconut, grape, **nutty**
26.58	1617	1624	MS, RI, Std	**Ethyl decanoate**	200	Fruit, pleasant, soap, sweet, **waxy**
28.07	1659	1661	MS, RI, Std	**Diethyl succinate**	500,000	Fabric, floral, fruit, lavender, potato, sweat
29.34	1693	1696	MS, RI	**Ethyl 3-hydroxyhexanoate**	265	**Citrus**, fruit, green, sweet
29.72	1715	1708	MS, RI	**Ethyl 9-decenoate**	100	Fruit, fatty
32.07	1773	1775	MS, RI, Std	**Ethyl benzeneacetate**	-	Fruit
34.52	1838	1837	MS, RI, Std	**Ethyl dodecanoate**	500	Fruit, soap, sweet
34.99	1857	1839	MS, RI, Std	**Ethyl salicylate**	-	Balsamic, cooling, **floral**, fruit, **spice**, sweet
35.58	1873	1870	MS, RI	**Benzyl butanoate**	-	**Floral**, fruit, jasmin, sweet,
35.87	1880	1883	MS, RI, Std	**Ethyl 2-phenylacetate**	250	**Floral**
66.75	2354	2358	MS, RI, Std	**Diethyl tartrate**	-	-
68.62	2420	2440	MS, RI, Std	**Ethyl succinate**	-	Fruit
70.49	2486	2499	MS, RI	**Ethyl citrate**	-	**Floral**
				Acids		
18.51	1425	1426	MS, RI, Std	**Acetic acid**	200,000	Pungent, vinegar, sour
23.52	1547	1557	MS, RI, Std	**2-Methylpropanoic acid**	200,000	Cheesy, fatty, phenolic, sweaty
25.97	1600	1607	MS, RI, Std	**Butanoic acid**	10,000	Buttery, cheesy, rancid, sweaty
27.56	1645	1647	MS, RI, Std	**3-Methylbutanoic acid**	3000	Cheesy, rancid, sweaty
34.49	1837	1840	MS, RI, Std	**Hexanoic acid**	3000	Cheesy, pungent, rancid, sweaty
36.92	1978	1981	MS, RI, Std	**2-Ethylhexanoic acid**	-	Cheesy
41.82	2098	2089	MS, RI, Std	**Octanoic acid**	10,000	Cheesy, fatty, fresh, moss
48.56	2321	2317	MS, RI, Std	**Decanoic acid**	15,000	Cheesy, fatty, soap
67.81	2392	2407	MS, RI	**Undecylic acid**	40	Oil

Table 1. *Cont.*

RT (min) [1]	KI_{calc} [2]	KI_{lit} [3]	ID [4]	Chemical Families	OTs (μg/L) [5]	Odor descriptors [6]
				Acetals		
4.99	1094	1096	MS, RI, Std	**1,1-Diethoxyethane**	1000	Liquorices, **nutty**, pungent, **wood**
21.86	1512	1525	MS, RI, Std	***Cis*-dioxane**	-	**Wood**
26.84	1642	1639	MS, RI, Std	***Cis*-dioxolane**	-	**Wood**
31.01	1755	1740	MS, RI, Std	***Trans*-dioxane**	-	**Wood**
				Furanic compounds		
18.03	1412	1434	MS, RI, Std	**2-Acetylfuran**	-	Balsamic-cinnamic, cereal, sweet, **toast**, **tobacco**
20.86	1465	1458	MS, RI, Std	**2-Furfural**	14,100	**Almond**, **caramel**, sweet, **wood**
22.45	1526	1524	MS, RI, Std	**1-(2-Furyl)-1-propanone**	-	Radish, **spice**
23.67	1550	1560	MS, RI, Std	**5-Methyl-2-furfural**	20,000	Acid, **almond**, **caramel**, coffee, **spice**, **toast**
26.01	1606	1606	MS, RI, Std	**Ethyl 2-furoate**	16,000	Balsamic, scorched tone, **vanilla**
68.15	2412	-	MS, RI, Std	**5-Ethoxymethyl-2-furfural**	6	Curry, **spice**
75.02	2501	2509	MS, RI, Std	**5-Hydroxymethyl-2-furfural**	10,000	**Almond**, cardboard, **nutty**
				Lactones		
25.71	1594	1595	MS, RI, Std	Butyrolactone	35,000	**Caramel**, coconut, cream, peach
29.01	1690	1694	MS, RI, Std	γ-Hexalactone	1600	Apricot, peach
36.65	1936	1933	MS, RI, Std	γ-Octalactone	400	**Caramel**, coconut, cream, fatty, herbaceous, **nutty**
42.75	2197	2185	MS, RI, Std	γ-Decalactone	88	Fruit, sweet
43.92	2218	2219	MS, RI	(Z)-Whiskylactone	67	**Caramel**, coconut, **nutty**, **toast**, **wood**
45.44	2267	2241	MS, RI, Std	γ-Dodecalactone	1000	Coconut, fruit, musk, sweet
				Volatile phenols		
41.63	2076	2080	MS, RI, Std	**2-Phenoxyethanol**	-	Alcoholic, **floral**, rose
45.13	2257	2250	MS, RI, Std	**Eugenol**	5	Balsamic, clove, herbaceous, **honey**, **spice**
77.06	2563	2561	MS, RI, Std	Vanillin	4	Sweet, **vanilla**
78.77	2620	2613	MS, RI, Std	Methyl vanillate	990	**Vanilla**

[1] Retention time (min); [2] Kovats index *n*-alkanes (C_8 to C_{20}) on a BP-20 capillary column; [3] Kovats index reported in literature for equivalent capillary column [21–30]; [4] Method of identification: MS, mass spectrum comparison using NITS library; RI: retention index in agreement with literature value; Std: confirmed by authentic standard; [5] Odor threshold determined in 10–12% *v/v* ethanol [6,31–36]; [6] Odor descriptors [6,11,37–45]; [7] Odor descriptors in bold are the potential aroma notes of Madeira wines; [8] VOCs in bold are common to all Madeira wines analyzed; [9] No information was found in literature; [10] TDN: 1,2-dihydro-1,1,6-trimethylnaphtalene.

2.1. Potential Impact Odorants of Madeira Wines

As observed in Table 2, young (3 to 5 years old) Madeira wines are characterized by freshness and fruitiness descriptors (e.g., citrus, floral, fruity), whereas the old (10 to 20 years old) Madeira wines are characterized by caramel, dried fruits, spice, toast and wood notes based on sensory analysis [6,46]. The information reported in Table 2 was obtained by a panelist of eleven expert judges belonging to different Madeira wine companies, such Madeira Wine Company, Barbeito, Borges, Justinos, and some employees from Instituto do Vinho Madeira, and also supported on previous studies [6,46].

Table 2. Madeira wines sensory analysis obtained from different grape varieties, types, and ages.

Madeira Wine Sensory Analysis		
Variety	**Younger wines (3 to 5 Years Old)**	**Older wines (10 to 20 Years Old)**
Malvasia	Almond, banana, citrus, cocoa, floral, tobacco, wood	Almond, caramel, dried fruits, spice, tobacco, toast, vanilla, wood
Bual	Almond, banana, cocoa, floral, tea	Almond, caramel, dried fruits, spice, tea, toast, wood
Sercial	Citrus, honey, mushroom, waxy	Dried fruits, honey, spice, toast, vanilla, wood
Verdelho	Banana, floral, honey, mushroom, spice	Dried fruits, ethereal, honey, spice, toast, wood
Tinta Negra	Citrus, ripe fruit, tea, wood	Caramel, dried fruits, spice, tea, toast, wood

A relationship-based approach consisting of two different nodes was built: (i) 15 Madeira wine aroma notes, and (ii) 82 VOCs that are known to explain each of these aroma notes (Figure 1).

The concentration and OT of each VOC is necessary to determine its contribution to overall Madeira wine aroma. In the current research, a semi-quantification was performed in order to establish a potential relationship between Madeira wines profiling and their odor descriptors with wines sensorial evaluation. As observed in Figure 1, different aroma notes were found for the same VOC, which could be influenced by odor the intensity evaluation, as well as VOCs concentration and nature of matrix analyzed. So, the resulting aroma relationship-based approach is too complex to achieve more information [6,11,37–45]. A projection of this relationship-based approach is the aroma system (Figure 2), in which two nodes (Madeira wine aroma notes) are linked if they share at least one aroma note. The color line represents the number of shared compounds.

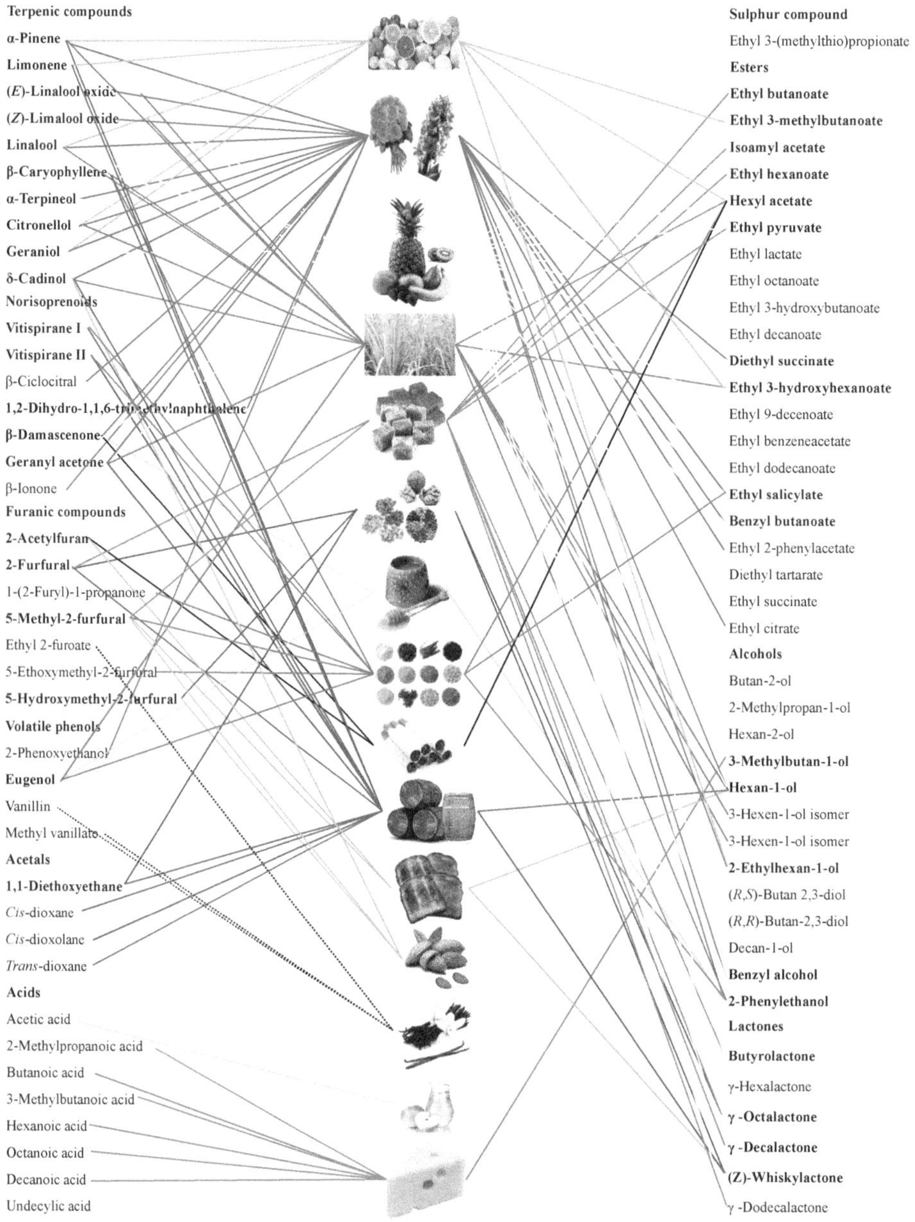

Figure 1. Madeira wine aroma notes (middle column), together with the chemical families (82 VOCs) that explain its aroma notes (left and right columns). Volatile compounds are shown in boldface if shared at least two or more aroma odors.

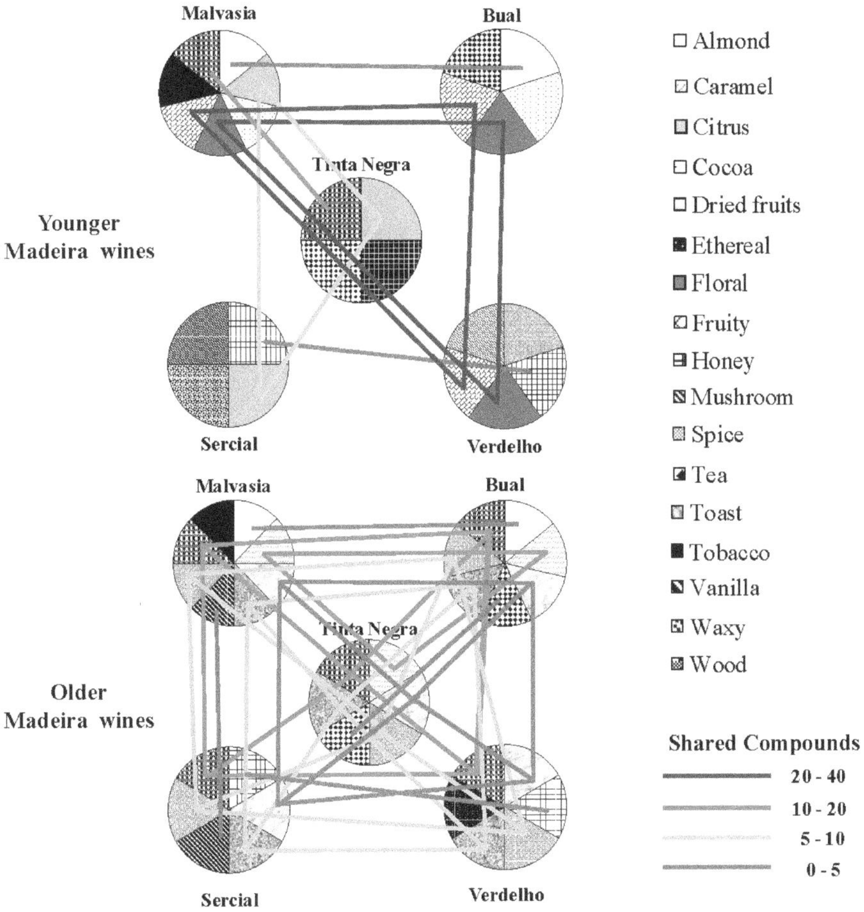

Figure 2. Aroma system. Each color represents an aroma note. The color line corresponds to the number of shared VOCs.

According to the obtained results, the grape variety seems to have a great impact in the sensorial properties of young Madeira wines, among several other parameters (e.g., vinification procedure). Specific aroma notes are linked to grape variety, as for example Malvasia and Bual grapes, used to produce sweet and medium sweet wines characterized by almond and cocoa odors. Dry and medium dry Madeira wines, obtained from Sercial and Verdelho grapes, are characterized by mushroom and honey notes. These grape varieties specific notes are shared, on average, by three VOCs. However, few aroma notes are connected between these grape varieties. Malvasia, Bual and Verdelho grapes are connected by flower and fruit odors and are shared, on average, by 31 VOCs, whereas the citrus odors linked to Malvasia, Tinta Negra and Sercial are shared by nine VOCs. In terms of Madeira aroma notes, young wines from Malvasia and Bual grapes are the most complex, contrarily to the observed for Tinta Negra, Sercial and Verdelho (Figure 2). For the oldest wines, it was observed that several aroma notes (e.g., dried fruit, spice, toast, wood) were present in all varieties under study. On average, eight VOCs that may explain these notes were shared by these varieties. Figure 2 shows that oldest wines, from the five varieties, presented higher aroma similarity than in youngest ones, which suggest the powerful impact of aging process on Madeira wine aroma.

2.1.1. Young Madeira Wines

Taking into account the OTs [6,31–36] and odor descriptor [6,11,37–45] reported in Table 1, as well as the relative concentration of VOCs (Tables S1 and S2), the citrus odor characteristic of Malvasia, Sercial and Tinta Negra wines (Table 2) could be explain by the presence of some terpenic compounds, such as α-pinene, limonene, linalool, citronellol, geraniol, and some esters, like hexyl acetate, ethyl 3-methylbutanoate and ethyl 3-hydroxyhexanoate, and 2-ethylhexan-1-ol (Table 3). All these varietal and fermentative compounds are present in Malvasia, Sercial and Tinta Negra wines at relative concentrations higher than their respective OT.

Table 3. Potential impact odorants of Madeira wine.

Odor Descriptor	Madeira Wines	Potential Odorant
Citrus	Malvasia, Sercial, TN	α-pinene, limonene, linalool, citronellol, geraniol, hexyl acetate, ethyl 3-methylbutanoate, ethyl 3-hydroxyhexanoate, 2-ethylhexan-1-ol
Floral	Malvasia, Bual, Verdelho	α-pinene, linalool, citronellol, geraniol, β-cyclocitral, TDN [1], β-damascenone, geranyl acetone, β-ionone, 1-hexanol, 2-phenylethyl alcohol
Waxy	Sercial	geraniol, geranyl acetone, ethyl octanoate, ethyl decanoate
Almond	Malvasia, Bual, Tinta Negra	δ-cadinol
Caramel	Malvasia, Bual	ethyl butanoate, ethyl hexanoate, ethyl pyruvate, (Z)-whiskylactone
Ethereal	Verdelho	ethyl lactate, ethyl pyruvate
Spice	Malvasia, Bual, Verdelho, Sercial, TN	hexyl acetate, 2-phenyethyl alcohol, 5-(ethoxymethyl)furfural, eugenol
Toast	Malvasia, Bual, Verdelho, Sercial, TN	(Z)-whiskylactone
Wood	Malvasia, Bual, Verdelho, Sercial, TN	δ-cadinol
Vanilla	Malvasia, Sercial	ethyl 2-furoate, vanillin, methyl vanillate

[1] TDN: 1,2-dihydro-1,1,6-trimethylnaphthalene; TN: Tinta Negra.

α-Pinene, linalool, citronellol, geraniol, β-cyclocitral, 1,2-dihydro-1,1,6-trimethylnaphthalene (TDN), β-damascenone, geranyl acetone, β-ionone, 1-hexanol and 2-phenylethyl alcohol are some varietal aroma compounds that could explain the floral odor related to Malvasia, Bual and Verdelho young wines (Table 3). By the other hand, linalool oxide and α-terpineol cannot explained the floral odors, since they are present in Malvasia, Bual and Verdelho wines at relative concentrations lower than their OT. Generally, the relative concentration of these varietal VOCs (e.g., α-pinene, linalool, citronellol) decreased during aging process (Tables S1 and S2), which could explain the absence of these odors in old wines. Some varietal compounds, linalool [6], β-damascenone [10] and TDN [47] have been reported as important odorants related to violet, exotic fruit and/or exotic floral descriptors of young wines. The waxy odor of Sercial young wine could be explained by the presence of some terpenic compounds (e.g., geraniol, geranyl acetone) and esters (e.g., ethyl octanoate, ethyl decanoate), since their relative concentration decreased slightly during aging process and present low OTs (Table 1). Finally, the almond odor of Malvasia and Bual young wines could be explained by the presence of δ-cadinol.

2.1.2. Old Madeira Wines

The caramel descriptor characteristic of older Malvasia, Bual and Tinta Negra wines suggests the presence of some esters (e.g., ethyl butanoate, ethyl hexanoate, ethyl pyruvate), furans (e.g., 2-furfural, 5-methyl-2-furfural), and some lactones (e.g., γ-butyrolactone, γ-octalactone, (Z)-whiskylactone). The relative concentration of furans and lactones increased during Madeira wines aging (Figure 3b).

Nevertheless, 2-furfural (OT = 14,100 μg/L), 5-methyl-2-furfural (20,000 μg/L), and γ-butyrolactone (OT = 35,000 μg/L) could not be used to explain the caramel descriptor since their relative concentration (Tables S1 and S2) was lower than their respective OTs (Table 1). Campo et al. [6] reported that furfural, 5-methylfurfural, 5-hydroxymethylfurfural and 5-ethoxymethylfurfural were not detected in the GC–O assays, in spite of the fact that these furanic compounds are quantitatively important, are not relevant to the aroma of Madeira wine. Moreover, phenylacetaldehyde, sotolon, (*Z*)-whiskylactone and some volatile phenols from wood are important odor active compounds in Madeira wines [6]. In the current study, from these three VOCs, only (*Z*)-whiskylactone was detected.

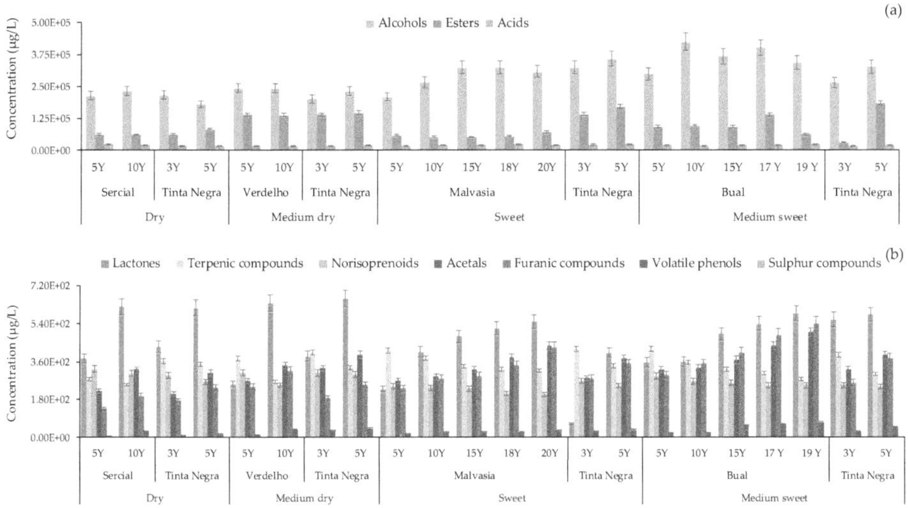

Figure 3. Total relative concentration (μg/L) of major (**a**) and minor (**b**) chemical families identified in Madeira wine.

Thus, based on the OTs, ethyl butanoate, γ-octalactone and (*Z*)-whiskylactone could be the VOCs responsible for the caramel descriptor characteristic of older Malvasia, Bual and Tinta Negra wines, since their relative concentrations (Tables S1 and S2) were higher than their OTs. The ethereal descriptor characteristic of Verdelho wines suggests the presence of ethyl lactate and ethyl pyruvate. Perhaps, these two VOCs were also presented in all Madeira wines analyzed, the relative concentration of ethyl pyruvate in Verdelho wines (on average 3.54 μg/L) was higher than the remaining Madeira wines (on average 1.49 μg/L).

The presence of hexyl acetate, 2-phenyethyl alcohol, 5-(ethoxymethyl) furfural and eugenol could explain the spicy notes characteristic of old Madeira wines, since their relative concentration was higher than their respective OTs (Table 1). Other VOCs that could explain the spicy notes were ethyl pyruvate and ethyl salicylate, however no information related to their OTs is available. 2-Acetylfuran, 5-methyl-2-furfural and (*Z*)-whiskey lactone could explain the toast odor, as their relative concentration increased remarkably during aging process (Tables S1 and S2). Nevertheless, taking into account the relative concentration and OTs, (*Z*)-whiskey lactone is the potential odorant responsible for the toast notes characteristic of older Madeira wines. The vanilla odor related to Malvasia and Sercial wines could be explained by the presence of ethyl 2-furoate, vanillin and methyl vanillate, since a remarkably increase on relative concentration was observed for vanillin and ethyl vanillate during aging process. Vanillin is the one of the VOCs that could explain the vanilla descriptor, since its relative concentration was higher than their OT (Table 1). δ-Cadinol, and acetals, like 1,1-diethoxyethane, *cis*-dioxane, *cis*-dioxolane and *trans*-dioxane could explain the wood descriptor characteristic of older Madeira wines, as their relative concentration slightly increased during aging process (Tables S1 and S2). In regards to the acetals, a little contribution to the sensorial properties of all Madeira wines was expected due to its high OT, and low relative concentration. However, in previous studies, 1,1-diethoxyethane [48] has been considered an important impact odorant to wines and liquors aromas [42,48], despite its higher OT. In the current study, 1,1-diethoxiethane was present in all Madeira wines analyzed at relative concentration lower than its OT.

Dried fruits notes (e.g., almond, coconut, nutty, peanut, walnut) characteristic of old Madeira wines could be explained mainly by the presence of 2-furfural, 5-hydromethyl-2-furfural, (*Z*)-whiskey lactone, γ-octalactone, γ-decalactone, γ-dodecalactone and 1,1-diethoxyethane, as their relative concentration increased remarkably during aging process (Tables S1 and S2). From these VOCs, only (*Z*)-whiskey lactone and γ-decalactone are present at relative concentration higher than its OTs (Table 1).

From a sensorial point of view, as can be observed in Figure 3, during the aging process the wine lost their freshness and fruitness odors related mainly to the presence of terpenic compounds (e.g., linalool oxide, linalool, α-terpeniol, geraniol), norisoprenoids (e.g., TDN, β-damascenone, geranyl acetone), and ethyl esters (e.g., ethyl 3-methylbutanoate, isoamyl acetate, ethyl 3-hydroxyhexanoate), as their relative concentration decreased during aging process (Tables S1 and S2). Otherwise, other descriptors arose such as caramel, dried fruit, spice, toast and wood, that suggests the formation of VOCs by Maillard reaction (e.g., furanic compounds), takes place at 50 °C being favored at pH 4–7 [49], and diffusion from the oak to wines (e.g., lactones, volatile phenols). As can be observed in Figure 3b, the relative concentration of these chemical families increased remarkably during wine aging, which could explain the aroma complexity of older Madeira wines.

3. Materials and Methods

3.1. Sampling

Twenty-two monovarietal Madeira wines from five *Vitis vinifera* L. grape varieties (one red—Tinta Negra, and four white wines from noble varieties—Malvasia, Bual, Sercial, and Verdelho), aged from 3 to 20 years old (Y) and matured in oak casks, were used in this study. Tinta Negra is the main grape variety harvested in Madeira Island (Portugal) representing more than 80% of the vineyards. The grapes were destemmed, crushed and 50 mg/L of sulfur dioxide (SO_2) was added. According to the age, the wines under study correspond to Vintage (a specific year of aged in casks, 17, 18, 19, and 20 years) and blended wines (B, an average aging period of 3, 5, 10, or 15 years). Four types of wine were used: Sweet (Malvasia, Tinta Negra, sugar content expressed as 96.1 to 150 g glucose per L), medium sweet (Bual, Tinta Negra, 80.4 to 96.1 g/L), dry (Sercial, Tinta Negra, 49.1 to 64.8 g/L), and medium dry (Verdelho, Tinta Negra, 64.8 to 80.4 g/L), and were aged in American oak casks. The ethanol content of the Madeira wines under study ranged from 18 to 19% (*v*/*v*).

3.2. Reagents and Standards

Sodium chloride (99.5%, foodstuff grade) and 4-methyl-2-pentanol (98%, internal standard) was purchased from Sigma Aldrich (Madrid, Spain), and ultrapure water was obtained from a Milli-Q system from Millipore (Milford, MA, USA). α-Pinene, linalool, α-terpeniol, citronellol, geraniol, β-ciclocitral, β-damascenone, geranyl acetone, β-ionone, butan-2-ol, hexan-2-ol, 3-methylbutan-1-ol, hexan-1-ol, (*E*)-3-hexen-1-ol, (*Z*)-3-hexen-1-ol, 2-ethylhexan-1-ol, decan-1-ol, benzyl alcohol, 2-phenethyl alcohol, ethyl butanoate, ethyl 3-methylbutanoate, isoamyl acetate, ethyl hexanoate, hexyl acetate, ethyl lactate, ethyl octanoate, ethyl 3-hydroxybutanoate, ethyl decanoate, ethyl benzeneacetate, ethyl dodecanoate, ethyl salicylate, ethyl 2-phenylacetate, ethyl succinate, acetic acid, butanoic acid, 3-methylbutanoic acid, hexanoic acid, 2-ethylhexanoic acid, octanoic acid, decanoic acid, 2-furfural, 1-(2-furyl)-1-propanone, 5-methyl-2-furfural, ethyl 2-furoate, 5-hydromethyl-2-furfural, butyrolactone, γ-hexalactone, γ-octalactone, γ-decalactone, γ-dodecalactone, 2-phenoxyethanol, eugenol, vanillin and methyl vanillate used for identification of target compounds were purchased from Sigma–Aldrich (Madrid, Spain), Acros Organics (Geel, Belgium), and Fluka (Buchs, Switzerland) with purity higher than 98%. The SPME holder for manual sampling and fiber were purchased from Supelco (Aldrich, Bellefonte, PA, USA). The SPME device included a fused silica fiber coating partially cross-linked with 50/30 μm divinylbenzene/carboxen/polydimethylsiloxane (DVB/CAR/PDMS). Prior to use, the SPME fiber was conditioned at 270 °C for 60 min in the GC injector, according to the manufacturer's recommendations. Then, the fiber was daily conditioned for 10 min at 250 °C. The *n*-alkane series analytical standard, C_8 to C_{20} straight-chain alkanes (concentration of 40 mg /L in *n*-hexane), used to determine the Kovat's index (KI) was supplied from Fluka (Buchs, Switzerland).

3.3. Sensory Analysis

A descriptive sensory analysis of Madeira wines samples used in this study was conducted by a panelist of eleven expert judges (seven females, four males) with an average age of 35 (± 5.1). The eleven members of the panel are winemakers belonging to different wine companies, such Madeira Wines, Barbeito, Borges, Justinos, and some employees from Instituto do Vinho Madeira. The panelists were trained over a period of 70 days to assess wine aroma using a "Le Nez du Vin" aroma kit (supplied by Ease Scent Company, Beijing, China). The "Le Nez du Vin" is an aroma kit composed by 54 vials, where each vial contains one typical aroma character in wine, such as blackcurrant, honey, caramel, coffee, chocolate, green pepper, smoke, wood, mint, among others. The training was carried out three times each week for 60–90 min. Each wine (30 mL) was presented to panelists in standard wine tasting glasses coded with three-digit numbers, covered with a Petri dish (to minimize the escape of VOCs), at 19–22 °C, in isolated booths under daylight-type lighting, with randomized presentation order. Cold water was used as palate cleansing. All wines were evaluated in triplicate in three formal sessions that were held on different days.

3.4. Headspace Solid-Phase Microextraction Tandem with Gas Chromatography-Mass Spectrometry (HS-SPME) methodology

The HS-SPME experimental parameters were previously established [15]. Briefly, aliquots of 4 mL of the wine sample were placed into an 8 mL glass vial. After the addition of 0.5 g of NaCl, 10 μL of 4-methyl-2-pentanol (internal standard, 250 μg/L) and stirring (0.5 × 0.1 mm bar) at 400 rpm, the vial was capped with a polytetrafluoroethylene (PTFE) septum and an aluminum cap (Chromacol, Hertfordshire, UK). The vial was placed in a thermostatted bath adjusted to 40.0 ± 0.1 °C for 5 min, and then the DVB/CAR/PDMS fiber was inserted into the headspace for 30 min. Three independent aliquots of each sample were analyzed in triplicate. Blanks, corresponding to the analysis of the coating fiber not submitted to any extraction procedure, were run between sets of three analysis.

3.5. GC–qMS Analysis for Madeira Wines Profiling

The GC–qMS methodology was based on a previous study [50]. After the extraction/concentration step, the SPME coating fiber containing the VOCs was manually introduced into the GC injection port at 250 °C (equipped with a glass liner, 0.75 mm I.D.) and kept for 7 min for desorption. The desorbed VOCs were separated in an Agilent Technologies 6890N Network gas chromatography equipped with a BP-20 fused silica capillary column (30 m × 0.25 mm I.D. × 0.25 μm film thickness) supplied by SGE (Darmstadt, Germany) connected to an Agilent 5973N quadrupole mass selective detector. Helium (Air Liquid, Portugal) was used as the carrier gas at a flow rate of 1.0 mL/min (column-head pressure: 12 psi). The injections were performed in the splitless mode (7 min). The GC oven temperature was programmed as follows: 45 °C (1 min) then ramped at 2 °C/min to 100 °C (3 min), 5 °C/min to 130 °C (5 min), and finally 2 °C/min to 220 °C (2 min). For the MS system, the temperatures of the transfer line, quadrupole and ionization source were 250, 150 and 230 °C, respectively; electron impact mass spectra were recorded at 70 eV and the ionization current was about 30 μA. The acquisitions were performed in full scan mode (30–300 *m/z*). The VOCs identification was achieved as follows: (*i*) comparison the GC retention times and mass spectra with those of the standard, when available; (*ii*) all mass spectra were also compared with the data system library (NIST, 2005 software, Mass Spectral Search Program v.2.0d; Nist 2005, Washington, DC); (*iii*) Kovat's index (KI) values were determined according to the van den Dool and Kratz equation [51]. For the KI determination, a C_8–C_{20} n-alkanes series was used, and the values were compared, when available, with values reported in the literature for similar columns [21–30].

The VOCs concentration was estimated, semi quantitatively, using the added amount of 4-methyl-2-pentanol (IS) according the following equation: VOCs concentration = (VOC GC peak area/IS GC peak area) × IS concentration. However, our main aim is regarding the relation between the varietal, fermentative and aging aroma compounds, and their aroma descriptors with the Madeira wine sensorial properties. This semi quantification approach was already performed in previous scientific studies [52,53].

4. Conclusions

This study represents the first detailed research about the Madeira wines volatile profiling and its association with odor descriptors. An in-depth relation among the varietal, fermentative and aging aroma compounds and their odor descriptors with the Madeira wine sensory analysis (described by a trained panelist) was established. The Madeira wine aroma notes, the VOCs and their aroma descriptors, showed the data complexity and the difficulty to get information. From the aroma system, it can be verified that grape variety is an important parameter that influences the sensorial properties of young Madeira wines, whereas the old wines are highly influenced by the aging process.

From a sensorial point of view, during the aging process the wine lost its freshness and fruitiness odors, and other descriptors arise such as caramel, dried fruit, spice, toast and wood, that suggests the formation of VOCs by Maillard reaction (e.g., furanic compounds), and diffusion from the oak to wines (e.g., lactones, volatile phenols). In addition, young Madeira wines obtained from Malvasia and Bual grape varieties are more complex than those obtained from Tinta Negra, Sercial, and Verdelho. This trend is not observed for the old Madeira wines since, independently of the grape variety used, their aroma notes are balanced which means that aging process has a higher impact on aroma rather than grape variety.

It is important to point out that a detailed database about volatile composition of Madeira wine and the correspondent aroma descriptors was built, which may be useful to improve information about the specific aroma of Madeira wine and will represent a powerful tool to help on winemaker decisions.

Supplementary Materials: The following are available online, Table S1: Relative concentration (μg/L) and relative standard deviation (%RSD) of VOCs identified by HS-SPME/GC-qMS in dry (Sercial and Tinta Negra) and medium dry (Verdelho and Tinta Negra) Madeira wines, Table S2: Relative concentration (μg/L) and relative standard deviation (RSD) of VOCs identified by HS-SPME/GC-qMS in sweet (Malvasia and Tinta Negra) and medium sweet (Bual and Tinta Negra) Madeira wines.

Author Contributions: R.P. performed the design of experiments, GC-MS, data analysis data analysis and manuscript preparation; C.S. performed the support to the experiments, GC-MS and data analysis. J.S.C. performed the manuscript preparation and editing.

Funding: This research was also funded by FCT-Fundação para a Ciência e a Tecnologia (project PEst-OE/QUI/UI0674/2013, CQM, Portuguese Government funds), and Associação Regional para o Desenvolvimento da Investigação Tecnologia e Inovação (ARDITI) through the project M1420-01-0145-FEDER-000005—Centro de Química da Madeira—CQM$^+$ (Madeira 14–20). The authors also acknowledge for PhD grant (SFRH/BD/97039/2013) to Catarina L. Silva

Acknowledgments: The authors acknowledge H.M. Borges Company for kindly providing the samples, and the information related to sensory properties of Madeira wines.

Conflicts of Interest: The authors declare no conflicts of interest.

References

1. Perestrelo, R.; Albuquerque, F.; Rocha, S.M.S.M.; Câmara, J.S.J.S. Distinctive characteristics of madeira wine regarding its traditional winemaking and modern analytical methodologies. *Adv. Food Nutr. Res.* **2011**, *63*, 207–249. [PubMed]
2. Perestrelo, R.; Petronilho, S.; Câmara, J.S.; Rocha, S.M. Comprehensive two-dimensional gas chromatography with time-of-flight mass spectrometry combined with solid phase microextraction as a powerful tool for quantification of ethyl carbamate in fortified wines. The case study of Madeira wine. *J. Chromatogr. A* **2010**, *1217*, 3441–3445. [CrossRef] [PubMed]
3. Sánchez-Palomo, E.; Trujillo, M.; Ruiz, A.G.; Viñas, M.A.G. Aroma profile of malbec red wines from La Mancha region: Chemical and sensory characterization. *Food Res. Int.* **2017**, *100*, 201–208. [CrossRef] [PubMed]
4. Lytra, G.; Tempere, S.; Le Floch, A.; de Revel, G.; Barbe, J.-C. Study of Sensory Interactions among Red Wine Fruity Esters in a Model Solution. *J. Agric. Food Chem.* **2013**, *61*, 8504–8513. [CrossRef] [PubMed]
5. Pereira, A.C.; Reis, M.S.; Saraiva, P.M.; Marques, J.C. Analysis and assessment of Madeira wine ageing over an extended time period through GC-MS and chemometric analysis. *Anal. Chim. Acta* **2010**, *660*, 8–21. [CrossRef] [PubMed]
6. Campo, E.; Ferreira, V.; Escudero, A.; Marqués, J.C.; Cacho, J. Quantitative gas chromatography-olfactometry and chemical quantitative study of the aroma of four Madeira wines. *Anal. Chim. Acta* **2006**, *563*, 180–187. [CrossRef]
7. Mendes, B.; Gonçalves, J.; Câmara, J.S. Effectiveness of high-throughput miniaturized sorbent- and solid phase microextraction techniques combined with gas chromatography-mass spectrometry analysis for a rapid screening of volatile and semi-volatile composition of wines—A comparative study. *Talanta* **2012**, *88*, 79–94. [CrossRef] [PubMed]
8. Pereira, A.C.; Reis, M.S.; Saraiva, P.M.; Marques, J.C. Madeira wine ageing prediction based on different analytical techniques: UV–vis, GC-MS, HPLC-DAD. *Chemom. Intell. Lab. Syst.* **2011**, *105*, 43–55. [CrossRef]
9. Pereira, V.; Cacho, J.; Marques, J.C. Volatile profile of Madeira wines submitted to traditional accelerated ageing. *Food Chem.* **2014**, *162*, 122–134. [CrossRef]
10. Câmara, J.S.; Alves, M.A.; Marques, J.C. Changes in volatile composition of Madeira wines during their oxidative ageing. *Anal. Chim. Acta* **2006**, *563*, 188–197. [CrossRef]
11. Perestrelo, R.; Fernandes, A.; Albuquerque, F.F.; Marques, J.C.; Câmara, J.S. Analytical characterization of the aroma of Tinta Negra Mole red wine: Identification of the main odorants compounds. *Anal. Chim. Acta* **2006**, *563*, 154–164. [CrossRef]
12. Câmara, J.S.; Marques, J.C.; Alves, M.A.; Silva Ferreira, A.C. 3-Hydroxy-4,5-dimethyl-2(5H)-furanone levels in fortified Madeira wines: Relationship to sugar content. *J. Agric. Food Chem.* **2004**, *52*, 6765–6769. [CrossRef]
13. Câmara, J.S.; Marques, J.C.; Alves, A.; Ferreira, A.C.S. Heterocyclic acetals in Madeira wines. *Anal. Bioanal. Chem.* **2003**, *375*, 1221–1224. [CrossRef] [PubMed]
14. Alves, R.F.; Nascimento, A.M.D.; Nogueira, J.M.F. Characterization of the aroma profile of Madeira wine by sorptive extraction techniques. *Anal. Chim. Acta* **2005**, *546*, 11–21. [CrossRef] [PubMed]
15. Perestrelo, R.; Barros, A.S.; Câmara, J.S.; Rocha, S.M. In-depth dearch focused on furans, lactones, volatile phenols, and acetals as potentialage markers of Madeira wines by comprehensive two-dimensional gas chromatography with time-of-flight mass spectrometry combined with solid phase microextraction. *J. Agric. Food Chem.* **2011**, *59*, 3186–3204. [CrossRef]
16. Mellon, F.A. CHROMATOGRAPHY|Combined Chromotography and Mass Spectrometry. *Encycl. Food Sci. Nutr.* **2003**, 1294–1301.
17. Câmara, J.S.; Alves, M.A.; Marques, J.C. Development of headspace solid-phase microextraction-gas chromatography–mass spectrometry methodology for analysis of terpenoids in Madeira wines. *Anal. Chim. Acta* **2006**, *555*, 191–200. [CrossRef]
18. Câmara, J.S.; Herbert, P.; Marques, J.C.; Alves, M.A. Varietal flavour compounds of four grape varieties producing Madeira wines. *Anal. Chim. Acta* **2004**, *513*, 203–207. [CrossRef]
19. Pereira, V.; Albuquerque, F.M.; Ferreira, A.C.; Cacho, J.; Marques, J.C. Evolution of 5-hydroxymethylfurfural (HMF) and furfural (F) in fortified wines submitted to overheating conditions. *Food Res. Int.* **2011**, *44*, 71–76. [CrossRef]

20. Pereira, V.; Leça, J.M.; Gaspar, J.M.; Pereira, A.C.; Marques, J.C. Rapid determination of sotolon in fortified wines using a miniaturized liquid-liquid extraction followed by LC-MS/MS analysis. *J. Anal. Methods Chem.* **2018**, *2018*, 1–7. [CrossRef]
21. Cejudo-Bastante, M.J.; Durán, E.; Castro, R.; Rodríguez-Dodero, M.C.; Natera, R.; García-Barroso, C. Study of the volatile composition and sensory characteristics of new Sherry vinegar-derived products by maceration with fruits. *LWT—Food Sci. Technol.* **2013**, *50*, 469–479. [CrossRef]
22. Lee, S.-J.; Noble, A.C. Characterization of odor-active compounds in Californian chardonnay wines using GC-olfactometry and GC-mass spectrometry. *J. Agric. Food Chem.* **2003**, *51*, 8036–8044. [CrossRef] [PubMed]
23. Spínola, V.; Perestrelo, R.; Câmara, J.S.; Castilho, P.C. Establishment of Monstera deliciosa fruit volatile metabolomic profile at different ripening stages using solid-phase microextraction combined with gas chromatography–mass spectrometry. *Food Res. Int.* **2015**, *67*, 409–417. [CrossRef]
24. Agnihotri, V.K.; Agarwal, S.G.; Dhar, P.L.; Thappa, R.K.; Kapahi, B.K.; Saxena, R.K.; Qazi, G.N. Essential oil composition of Mentha pulegium L. growing wild in the north-western Himalayas India. *Flavour Fragr. J.* **2005**, *20*, 607–610. [CrossRef]
25. Choi, H.-S. Lipolytic effects of citrus peel oils and their components. *J. Agric. Food Chem.* **2006**, *54*, 3254–3258. [CrossRef]
26. Mallia, S.; Fernández-García, E.; Olivier Bosset, J. Comparison of purge and trap and solid phase microextraction techniques for studying the volatile aroma compounds of three European PDO hard cheeses. *Int. Dairy J.* **2005**, *15*, 741–758. [CrossRef]
27. Gürbüz, O.; Rouseff, J.M.; Rouseff, R.L. Comparison of aroma volatiles in commercial Merlot and Cabernet Sauvignon wines using gas chromatography−olfactometry and gas chromatography−mass spectrometry. *J. Agric. Food Chem.* **2006**, *54*, 3990–3996. [CrossRef]
28. Kim, T.H.; Shin, J.H.; Baek, H.H.; Lee, H.J. Volatile flavour compounds in suspension culture ofAgastache rugosa Kuntze (Korean mint). *J. Sci. Food Agric.* **2001**, *81*, 569–575. [CrossRef]
29. Culleré, L.; Escudero, A.; Cacho, J.; Ferreira, V. Gas chromatography-olfactometry and chemical quantitative study of the aroma of six premium quality Spanish aged red wines. *J. Agric. Food Chem.* **2004**, *52*, 1653–1660. [CrossRef]
30. Zhao, Y.; Xu, Y.; Li, J.; Fan, W.; Jiang, W. Profile of volatile compounds in 11 brandies by headspace solid-phase microextraction followed by gas chromatography-mass spectrometry. *J. Food Sci.* **2009**, *74*, C90–C99. [CrossRef]
31. Bertagnolli, S.M.M.; Bernardi, G.; Donadel, J.Z.; Fogaça, A.D.O.; Wagner, R.; Penna, N.G. Natural sparkling guava wine: Volatile and physicochemical characterization. *Ciência Rural* **2017**, *47*. [CrossRef]
32. Fazzalari, F.A. *Compilation of odor and taste threshold values data*; American Society for Testing and Materials: West Conshohocken, PA, USA, 1978.
33. Ferreira, V.; Aznar, M.; López, R.; Cacho, J. Quantitative gas chromatography-olfactometry carried out at different dilutions of an extract. Key differences in the odor profiles of four high-quality Spanish aged red wines. *J. Agric. Food Chem.* **2001**, *49*, 4818–4824. [CrossRef] [PubMed]
34. Ferreira, V.; López, R.; Cacho, J. Quantitative determination of the odorants of young red wines from different grape varieties. *J. Sci. Food Agric.* **2000**, *80*, 1659–1667. [CrossRef]
35. Guth, H. Quantitation and sensory studies of character impact odorants of different white wine varieties. *J. Agric. Food Chem.* **1997**, *45*, 3027–3032. [CrossRef]
36. Peinado, R.A.; Moreno, J.A.; Muñoz, D.; Medina, M.; Moreno, J. Gas chromatographic quantification of major volatile compounds and polyols in wine by direct injection. *J. Agric. Food Chem.* **2004**, *52*, 6389–6393. [CrossRef]
37. Ribéreau-Gayon, P.; Glories, Y.; Maujean, A.; Dubourdieu, D. *Handbook of Enology, The Chemistry of Wine: Stabilization and Treatments: Second Edition*; John Wiley & Sons, Ltd.: Chichester, UK, 2006; Volume 2, ISBN 9780470010396.
38. Acree, T.; Arn, H. Flavornet Home Page. Available online: http://www.flavornet.org/ (accessed on 14 February 2019).
39. Díaz-Maroto, M.C.; Guchu, E.; Castro-Vázquez, L.; de Torres, C.; Pérez-Coello, M.S. Aroma-active compounds of American, French, Hungarian and Russian oak woods, studied by GC–MS and GC–O. *Flavour Fragr. J.* **2008**, *23*, 93–98. [CrossRef]

40. Ducruet, V.; Fournier, N.; Saillard, P.; Feigenbaum, A.; Guichard, E. Influence of packaging on the aroma stability of strawberry syrup during shelf life. *J. Agric. Food Chem.* **2001**, *49*, 2290–2297. [CrossRef] [PubMed]
41. El-Sayed, A.M. The Pherobase: Database of insect pheromones and semiochemicals. Available online: http://www.pherobase.com (accessed on 7 July 2018).
42. Fan, W.; Qian, M.C. Characterization of aroma compounds of Chinese "Wuliangye" and "Jiannanchun" liquors by aroma extract dilution analysis. *J. Agric. Food Chem.* **2006**, *54*, 2695–2704. [CrossRef]
43. Genovese, A.; Gambuti, A.; Piombino, P.; Moio, L. Sensory properties and aroma compounds of sweet Fiano wine. *Food Chem.* **2007**, *103*, 1228–1236. [CrossRef]
44. Giri, A.; Osako, K.; Okamoto, A.; Ohshima, T. Olfactometric characterization of aroma active compounds in fermented fish paste in comparison with fish sauce, fermented soy paste and sauce products. *Food Res. Int.* **2010**, *43*, 1027–1040. [CrossRef]
45. Leffingwell, D.; Leffingwell, J.C. Odor detection thresholds of GRAS flavor chemicals. Available online: http://www.leffingwell.com (accessed on 1 July 2018).
46. Silva, H.O.; Guedes De Pinho, P.; Machado, B.P.; Hogg, T.; Marques, J.C.; Câmara, J.S.; Albuquerque, F.; Silva Ferreira, A.C. Impact of forced-aging process on Madeira wine flavor. *J. Agric. Food Chem.* **2008**, *56*, 11989–11996. [CrossRef] [PubMed]
47. Sacks, G.L.; Gates, M.J.; Ferry, F.X.; Lavin, E.H.; Kurtz, A.J.; Acree, T.E. Sensory threshold of 1,1,6-trimethyl-1,2-dihydronaphthalene (TDN) and concentrations in young Riesling and Non-Riesling wines. *J. Agric. Food Chem.* **2012**, *60*, 2998–3004. [CrossRef] [PubMed]
48. Moreno, J.A.; Zea, L.; Moyano, L.; Medina, M. Aroma compounds as markers of the changes in sherry wines subjected to biological ageing. *Food Control.* **2005**, *16*, 333–338. [CrossRef]
49. Pereira, V.; Santos, M.; Cacho, J.; Marques, J.C. Assessment of the development of browning, antioxidant activity and volatile organic compounds in thermally processed sugar model wines. *LWT* **2017**, *75*, 719–726. [CrossRef]
50. Perestrelo, R.; Rodriguez, E.; Câmara, J.S.J.S. Impact of storage time and temperature on furanic derivatives formation in wines using microextraction by packed sorbent tandem with ultrahigh pressure liquid chromatography. *LWT—Food Sci. Technol.* **2017**, *76*, 40–47. [CrossRef]
51. van Den Dool, H.; Dec. Kratz, P. A generalization of the retention index system including linear temperature programmed gas—liquid partition chromatography. *J. Chromatogr. A* **1963**, *11*, 463–471. [CrossRef]
52. Perestrelo, R.; Silva, C.; Silva, P.; Câmara, J.S. Unraveling Vitis vinifera L. grape maturity markers based on integration of terpenic pattern and chemometric methods. *Microchem. J.* **2018**, *142*, 367–376. [CrossRef]
53. Kang, W.; Xu, Y.; Qin, L.; Wang, Y. Effects of different β-D-glycosidases on bound aroma compounds in Muscat grape determined by HS-SPME and GC-MS. *J. Inst. Brew.* **2010**, *116*, 70–77. [CrossRef]

Sample Availability: Samples of the compounds are not available from the authors.

molecules

MDPI

Article

Development and Evaluation of a HS-SPME GC-MS Method for Determining the Retention of Volatile Phenols by Cyclodextrin in Model Wine

Chao Dang [1,2], Kerry L. Wilkinson [1,2,*], Vladimir Jiranek [1,2] and Dennis K. Taylor [1,2]

1 The University of Adelaide, School of Agriculture, Food and Wine, PMB 1, Glen Osmond, SA 5064, Australia; chao.dang@adelaide.edu.au (C.D.); vladimir.jiranek@adelaide.edu.au (V.J.); dennis.taylor@adelaide.edu.au (D.K.T.)

2 The Australian Research Council Training Centre for Innovative Wine Production, PMB 1, Glen Osmond, SA 5064, Australia

* Correspondence: kerry.wilkinson@adelaide.edu.au; Tel.: +61-8-8313-7360

Academic Editors: Rosa Maria de Sá Perestrelo and José Sousa Câmara

Received: 25 August 2019; Accepted: 20 September 2019; Published: 21 September 2019

Abstract: Volatile phenols exist in wine and can be markers for *Brettanomyces* and smoke taint off-odors. Cyclodextrins (CDs) are found to be capable of forming inclusion complexes with volatile phenols. Cross peaks on 2D ^{1}H ROESY nuclear magnetic resonance (NMR) spectra demonstrated inclusion of volatile phenols in the β-CD cavity, while difference tests confirmed this resulted in a perceptible reduction of their sensory impact. However, a conventional headspace solid phase microextraction (HS-SPME) method using an isotopically labelled normalizing standard failed to quantify the residual volatile phenols by gas chromatography-mass spectrometry (GC-MS) because of inclusion of the standard by the CDs. A new method involving an additional liquid phase was developed and validated for quantitation of volatile phenols in the presence of CDs. The retention of eight volatile phenols by α-, β-, and γ-CD was subsequently studied.

Keywords: *Brettanomyces*; cyclodextrins; gas chromatography-mass spectrometry; nuclear magnetic resonance; smoke taint; volatile phenols; wine

1. Introduction

Volatile phenols are an important group of wine aroma compounds. Some volatile phenols, for example guaiacol, 4-methylguaiacol, vanillin, and eugenol, are routinely identified in wines aged in oak barrels, as a consequence of thermal degradation of lignin during the toasting process of cooperage [1]. These volatile phenols contribute to the smoky, vanilla and clove characters often associated with oak maturation [2,3]. However, volatile phenols are also responsible for certain off-odors in wine. *Brettanomyces* and/or *Dekkera* spoilage can result in the accumulation of 4-ethylguaiacol and 4-ethylphenol in wine, which at elevated concentrations can impart undesirable barnyard, sweaty, medicinal, and/or horsy notes [4]. Guaiacols, cresols, and syringols have been identified as markers of smoke taint, i.e., the objectionable smoky, ashy character observed in wines made from grapes exposed to bushfire smoke [5–7].

The wine industry has long sought strategies for mitigating off-odors, including those attributable to the presence of volatile phenols. Most amelioration strategies have involved the addition of sorptive materials such as yeast lees [8,9], activated carbon [10], and polyvinylpolypyrrolidone [11] to remove taint compounds from wine. However, these materials can also bind to volatiles responsible for desirable wine aromas and flavors. Reverse osmosis fractionation of wine prior to solid phase adsorption treatment has been used to achieve more selective removal of taint compounds [12,13]; while novel sorbents including esterified cellulose [14], polyaniline-based compounds [15] and molecularly imprinted polymers [16,17] have also been evaluated for the amelioration of taint because of the presence of volatile phenols in wine.

Cyclodextrins (CDs) are cyclic oligosaccharides comprising α-1,4-linked glucose units, the most common being α-CD, β-CD, and γ-CD, which comprise 6, 7, and 8 glucose units, respectively [18] (Figure S1). The spatial arrangement of sugars gives CDs a characteristic bucket shape, whereby the hydrophilic outer surface affords water solubility, while the hydrophobic inner cavity enables the formation of host-guest inclusion complexes with various molecules, including volatile phenols [18,19] (Figure S2). Numerous industries have exploited the encapsulation of volatile compounds by CDs, including those involved in the production of foods, beverages, and cosmetics, to stabilize, preserve, and/or mask aromas, flavors, and fragrances [20–23]. However, to date, there are few studies concerning the use of CDs in winemaking. The potential for β-CD to reduce the intensity of off-odors associated with *Brettanomyces* spoilage of red wine has been demonstrated [24]. β-CD has also been used to extract stilbenes, flavonols, and flavan-3-ols from grapes and pomace [25]. A key aim of the current study is therefore to determine to what extent volatile phenols associated with *Brettanomyces* spoilage and smoke taint can form inclusion complexes with CDs, so as to mitigate their impact on wine sensory properties. In order to achieve this aim, the concentration of volatile phenols should be determined before and after the addition of CD to wine. Headspace solid-phase microextraction (HS-SPME) has been shown to be a fast and effective sampling method for gas chromatography-mass spectrometry (GC-MS) analysis and it has been used extensively for the determination of volatile compounds in wine [26,27], including volatile phenols [28]. However, quantitative analysis relies on the addition of an appropriate normalizing standard, for example an isotopically labelled analogue in the case of stable isotope dilution assays [29], and the standards are equally subject to treatments on the sample mixture, such as the addition of CDs. Whereas conventional HS-SPME employs a three-phase extraction system, comprising the sample, its headspace, and the SPME fiber, in the current study, an additional liquid phase was introduced to overcome interactions between CDs and normalizing standards. This was achieved by inserting a glass ampoule containing the internal standard solution into the headspace vial, prior to analysis. This study describes the development and validation of a novel four-phase HS-SPME GC-MS method for determining the retention of volatile phenols by CD in model wine.

2. Results and Discussion

2.1. Traditional (Three-Phase) HS-SPME GC-MS, NMR, and Sensory Analysis

The conventional three-phase HS-SPME GC-MS method yielded excellent repeatability and linearity for quantification of guaiacol, 4-methylguaiacol, and 4-ethylphenol in model wine (Table S1). In the CD treatment assay, 25 g/L of α-CD, β-CD, and γ-CD were dissolved in the mixture before the internal standard was added. The relative peak area (RPA) of treatment groups and the controls were compared, but no significant differences in volatile phenol levels were observed for any of the CD treatments (Table 1). These results contradicted previous reports that β-CD significantly reduced the sensory perception of 4-ethylphenol in wine [24]. It was speculated that the CDs may have formed an inclusion complex with both the volatile phenols and the internal standards, equalizing changes in the relative response of both samples and standards following CD addition. In agreement with previously published work [27–30], the absolute peak areas were not reproducible, particularly when different fibers were used, and thus were not suitable for use in quantification.

To establish whether or not binding had occurred between the CDs and volatile phenols, NMR analyses were carried out on the mixture of β-CD and several volatile phenols (Figure 1). Cross peaks arising from the nuclear Overhauser effect (NOE) were observed between the protons in the β-CD cavity and volatile phenol protons, confirming the close spatial arrangement of these protons in an inclusion complex. This was further supported by sensory analysis, with 24 and 20 panelists (of 38) perceiving a difference in the "smoke taint" and "*Brettanomyces*" brackets, respectively, suggesting significant changes in the volatile phenol levels in the headspace after β-CD treatment. These results confirmed that the three-phase headspace SPME method, involving addition of the internal standard to the sample containing CDs, was not suitable for quantitative analysis.

Table 1. Residual volatile phenol levels (as relative peak area) following addition of α-, β-, and γ-CDs to model wine, based on conventional three-phase headspace solid phase microextraction (HS-SPME) gas chromatography-mass spectrometry (GC-MS) analysis.

	Guaiacol	4-Methylguaiacol	4-Ethylphenol
Control	0.99 ± 0.02	1.00 ± 0.01	1.22 ± 0.02
α-CD	0.99 ± 0.02 (100%)	1.00 ± 0.02 (100%)	1.23 ± 0.02 (100.8%)
β-CD	0.99 ± 0.01 (100%)	1.00 ± 0.01 (100%)	1.29 ± 0.02 (105.7%)
γ-CD	0.99 ± 0.02 (100%)	1.00 ± 0.02 (100%)	1.24 ± 0.03 (101.6%)

Values are means of three replicates ± standard error (and percentage of control). Values within columns were not significantly different (one-way ANOVA, $p = 0.05$).

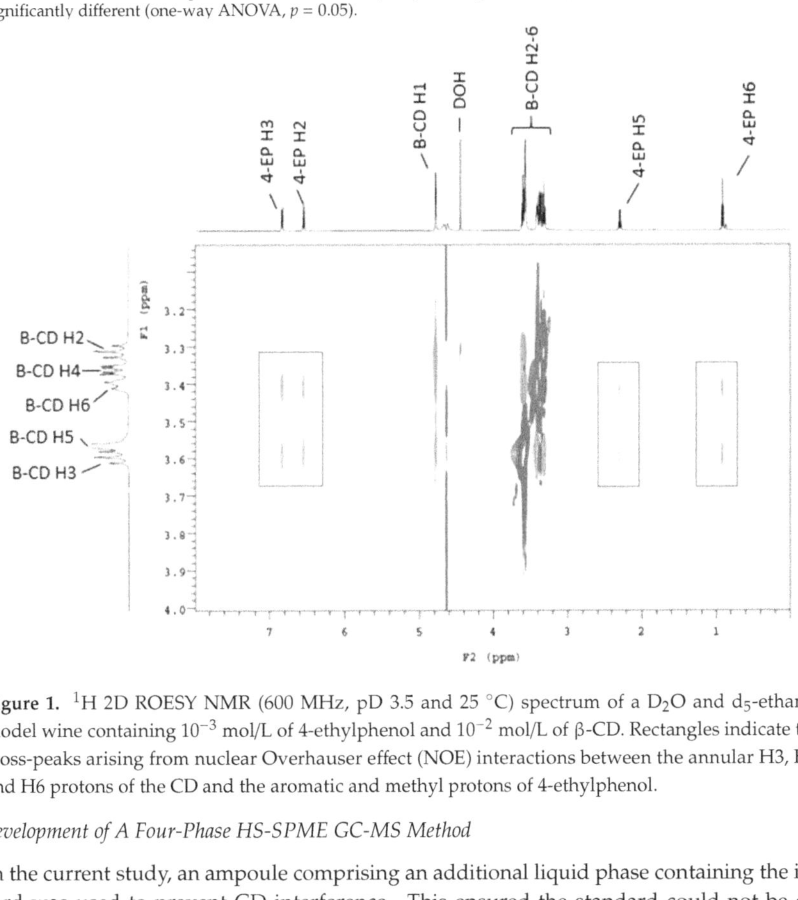

Figure 1. ^1H 2D ROESY NMR (600 MHz, pD 3.5 and 25 °C) spectrum of a D_2O and d_5-ethanol model wine containing 10^{-3} mol/L of 4-ethylphenol and 10^{-2} mol/L of β-CD. Rectangles indicate the cross-peaks arising from nuclear Overhauser effect (NOE) interactions between the annular H3, H5, and H6 protons of the CD and the aromatic and methyl protons of 4-ethylphenol.

2.2. Development of A Four-Phase HS-SPME GC-MS Method

In the current study, an ampoule comprising an additional liquid phase containing the internal standard was used to prevent CD interference. This ensured the standard could not be directly encapsulated by the CDs, but this modification significantly affected the kinetics of the existing SPME method. Conventionally, quantitative headspace SPME method is performed when a partition equilibrium of the target compound is achieved between three phases, namely the sample matrix, the headspace, and the fiber. The extraction of a given compound can then be expressed as:

$$C_0{*}V_s = C_s{*}V_s + C_h{*}V_h + C_f{*}V_f \quad (1)$$

where C_0 is the original concentration of the compound in the sample, C_s is the residual concentration remaining in the sample, Vs is the volume of the sample (liquid phase), C_h is the concentration of the compound in the headspace, V_h is the volume of the headspace (gas phase), C_f is the concentration of the compound on the SPME fiber coating, and V_f is the volume of the fiber (solid phase).

There are two equilibria in this process, i.e., the equilibrium between the sample and the headspace (K_1) and that between the headspace and the fiber coating (K_2). The equilibrium constants, i.e., Henry's Law Constants, K_1 and K_2, are expressed as: $K_1 = C_s/C_h$ and $K_2 = C_h/C_f$, where the equilibration time is longer than ideal, good precision can be achieved, provided extraction conditions such as temperature, fiber penetration, and agitation are well controlled [31]. In the current study, the equilibrium is more complex, comprising distribution of volatile compounds in four phases during extraction, because of the presence of an additional liquid phase:

$$C_0{*}V_s = C_s{*}V_s + C_i{*}V_i + C_h{*}V_h + C_f{*}V_f \quad (2)$$

where C_i is the concentration of the internal standard in the additional liquid phase, and V_i is the volume of the additional liquid phase. An additional equilibrium constant, K_3, exists for partitioning between the headspace and the additional liquid phase: $K_3 = C_h/C_i$.

In the current study, it is hypothesized that the sample mixture containing the CDs would not meaningfully interfere with the volatile phenols present in the headspace because of the relatively short extraction time, so $C_i{*}V_i$ was considered to be zero. The method development employed in this study did not focus on modelling the overall process, but rather the practicality of the process in determining the retention of volatile phenols by CDs.

2.2.1. Influence of Agitation

The RPAs obtained for some samples changed significantly with time, before more constant levels were achieved. With the same volume of internal standard, agitated samples yielded significantly higher RPAs than non-agitated samples, particularly for lower equilibration times (Figure 2). To further investigate, absolute peak areas for *m/z* 124 (guaiacol) and *m/z* 127 (d_3-guaiacol) were compared. Despite being unable to quantify the compounds, the absolute peak areas were used to establish a hypothesis, based on samples being analyzed in triplicate, using relatively new fibers (i.e., <50 injections), with no sign of degradation. The response of *m/z* 124 was generally higher in agitated samples than in non-agitated samples, provided the same volume of internal standard was used. In contrast, the opposite was observed for the response of *m/z* 127 (data not shown). To provide an explanation, the auto-sampler's agitation process was evaluated, and it was found that agitation had a variable effect on both the sample and the internal standard solution. At 250 rpm the agitator moved the headspace vial in a horizontal circular trajectory, with the inserted ampoule spinning within the vial. This caused the ampoule to sit at an angle in the vial, which impacted the relative abundance of *m/z* 124 in the headspace of agitated samples. According to Dalton's Law, the total pressure in a gas phase equals the sum of pressure of each individual component. In the current case, the headspace pressure in the vial is comprised of the vapor pressure of both the sample and the internal standard solution. The distribution of each volatile component is defined by its Henry's Law constant. In the concept of HS-SPME, Pawliszyn indicated that Henry's Law constant is only dependent on the system temperature and the liquid phase matrix [31]. Considering the sample vials were left unagitated prior to extraction, it can be concluded that agitation-induced partial pressure differences during extraction disrupted the headspace pressure distribution. However, this disruption doesn't alter the Henry's Law constant of any of the volatile compounds, or the final equilibrium, and it becomes less effective as the system approaches equilibrium. As a consequence, samples were extracted without agitation in the final HS-SPME method developed in this study.

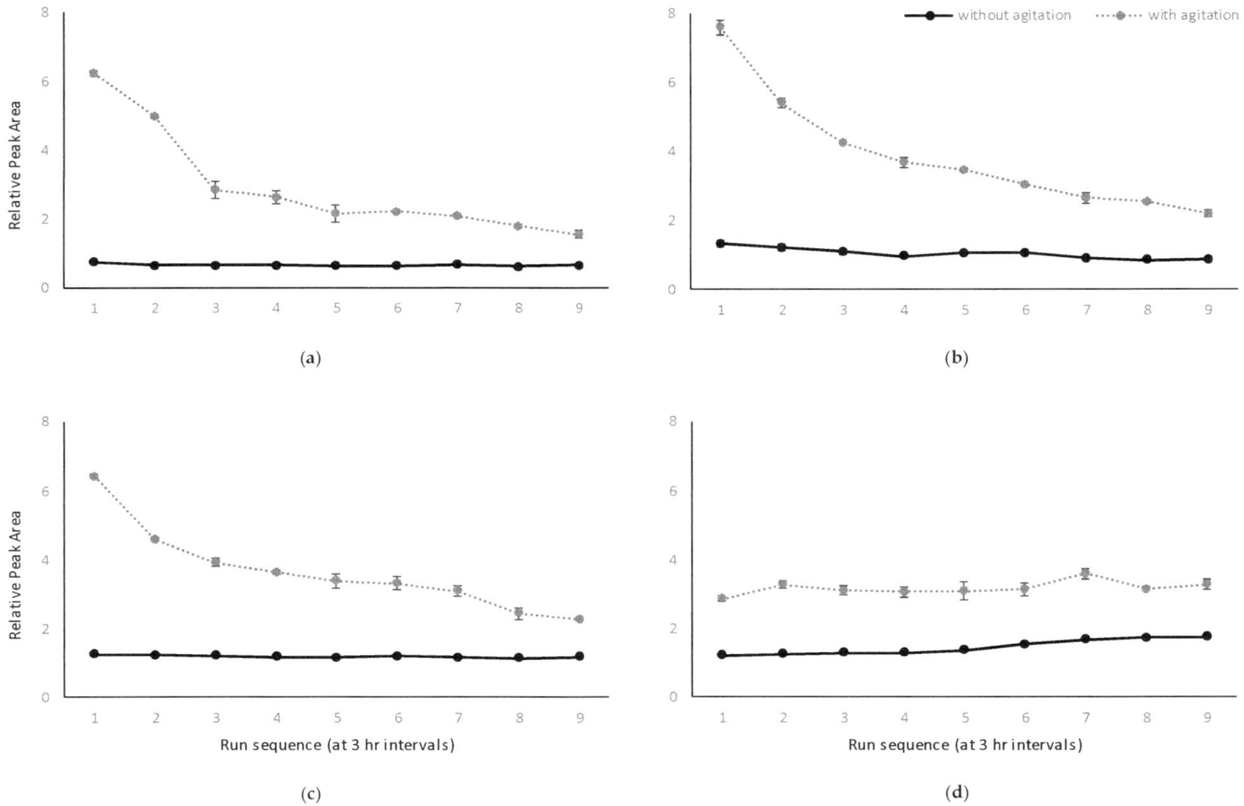

Figure 2. Effect of agitation during equilibration and internal standard volume, (**a**) 2 mL, (**b**) 1 mL, (**c**) 0.5 mL, and (**d**) 0.1 mL, on the relative peak area of guaiacol over time (i.e., expressed as run sequence at 3-hr intervals after the SPME vial was capped). Values are means of three replicates ± standard error.

2.2.2. Influence of Volume of Internal Standard Solution

According to Henry's Law Constant, equation $K_1 = C_s/C_h$ can be rearranged to give: $C_h = C_0/(K_1 + \beta)$; where β is the phase ratio between the headspace and the liquid phase of the sample. In the recently updated Henry's Constant Compilation [32], the value of K for guaiacol partitioning between water and air is around 2.2×10^4 at 25 °C (converted from reported H_{cp} mol/m^3 Pa). In the current study, the phase ratio (β) between the additional liquid phase and the headspace phase ranged from 6 to 139, which is insignificant when added to K (the volume of the glass material of the ampoule was deemed negligible). It can be inferred that the concentration of internal standard in the headspace at equilibrium would be within similar ranges for the various internal standard volumes used. It was concluded that agitation disrupted the equilibrating process, albeit only small deviations were observed in the RPA of agitated samples when the internal standard volume was 0.1 mL (Figure 2). This indicated that the system approached equilibrium sooner with smaller internal standard volumes. As such, lower volumes of internal standard were used in the new HS-SPME method. Taking into account the possible depletion of the deuterium labels in the internal standard [33], 0.5 mL was chosen as the functional volume for the internal standard.

2.2.3. Influence of Extraction Temperature, Extraction Time, and Internal Standard Concentration

Once the key analytical parameters had been optimized, several other factors, i.e., extraction temperature, duration, and internal standard concentration, were evaluated. Increasing RPAs were observed for volatile phenols when extraction temperature was increased from 35 to 80 °C (Figure 3). As mentioned above, Henry's Law constant (K) is temperature dependent, so K decreased as the extraction temperature increased for most volatile phenols [32]. Wieland and colleagues reported a 100-fold decrease in Henry's law constant for guaiacol, when the temperature increased from 35 to 80 °C [34]. The phase ratio (β) for the internal standard and sample was 27 and 2.25, respectively (for 0.5 mL of internal standard). According to $C_h = C_0/(K + \beta)$, with decreasing K, C_h will have greater increases at low β values. Three things need to be taken into consideration when choosing extraction temperature: experimental sensitivity; the stability of CD complexation; and the potential for volatile compounds present in the headspace to re-dissolve in either of the liquid phases (i.e., the sample or the internal standard solution). In the present study, a temperature of 35 °C was therefore chosen as a trade-off.

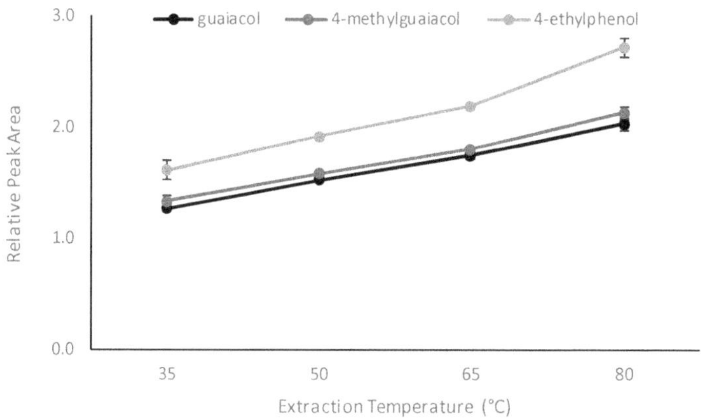

(a)

Figure 3. *Cont.*

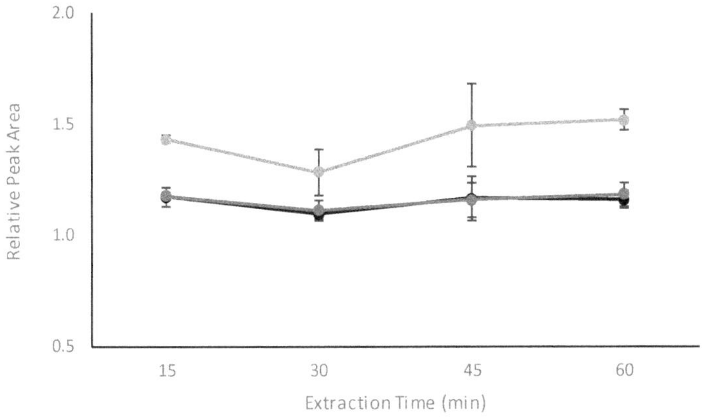

(**b**)

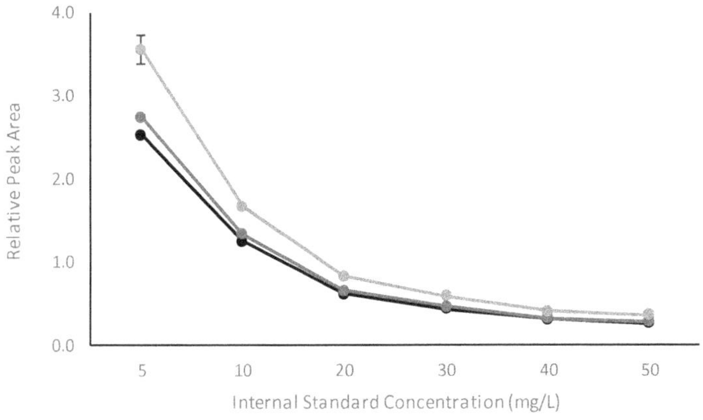

(**c**)

Figure 3. Effect of (**a**) extraction temperature, (**b**) extraction time, and (**c**) internal standard concentration on the relative peak area of guaiacol, 4-methylguaiacol, and 4-ethylphenol. Values are means of three replicates ± standard error (but some standard errors are obscured by symbols).

Since the equilibrium problem was resolved by avoiding agitation and minimizing the volume of internal standard, optimization of the extraction time largely addressed the analytical sensitivity. The RPA for volatile phenols did not show significant differences between extraction times, ranging from 15 to 60 min (Figure 3). Accordingly, 15 min was chosen as the extraction time for the new HS-SPME method.

In terms of internal standard concentration, according to $C_h = C_0/(K + \beta)$ and $RPA = A_s/A_i$, it is expected that the RPA will follow a rational function with linear increases in concentration of the internal standard. Experimental data supported this notion (Figure 3). Indeed, in quantification studies using an isotopically labelled standard, the concentration is normally irrelevant to the analysis of compounds of interest, except where the standard is used for calibration [35]. The concern with choosing a higher concentration of ISTD is the adsorption capacity of the SPME fiber and competition for absorption between the standard and the compounds of interest. In the current study, 10 mg/L gave a RPA range close to 1.0.

2.3. *Experimental Conditions for the Four-Phase HS-SPME GC-MS Method*

The final method involved a 6-mL aliquot of model wine solution spiked with volatile phenols (at 1 mg/L) being placed into a 20-mL headspace vial. A 0.5-mL aliquot of internal standard solution (10 mg/L) was added to a 2-mL glass ampoule, which was then inserted into the SPME vial. The vial was incubated for 10 min at 35 °C before extraction with the SPME fiber for 15 min. No agitation was employed during extraction. The new HS-SPME GC MS method gave excellent linearity and repeatability (Table 2). A calibration function was constructed for guaiacol, 4-methylguaiacol, and 4-ethylphenol ranging from 0.25 to 2 mg/L, and also gave excellent linearity, with an R^2 value ≥ 0.9956.

Table 2. Validation of the four-phase HS-SPME GC-MS method.

	Guaiacol		**4-Methylguaiacol**		**4-Ethylphenol**	
	RPA	**CV (%)**	**RPA**	**CV (%)**	**RPA**	**CV (%)**
0.25 mg/L	0.35	2.7	0.38	2.5	0.48	1.7
0.5 mg/L	0.63 (0.65)	3.1 (4.5)	0.68 (0.69)	2.2 (5.8)	0.86 (0.90)	5.4 (8.9)
1.0 mg/L	1.12	7.5	1.16	8.9	1.50	9.3
1.25 mg/L	1.61 (1.53)	0.6 (5.4)	1.74 (1.58)	0.8 (5.8)	2.27 (1.99)	6.5 (7.0)
1.75 mg/L	2.12 (2.10)	0.1 (2.2)	2.27 (2.17)	0.2 (3.1)	3.01 (2.84)	6.6 (6.1)
2.0 mg/L	2.31	0.7	2.36	0.8	2.97	2.4

Values are means of three replicates (and repeat analyses performed after 1 month). RPA = relative peak area; CV = coefficient of variation.

2.4. *Retention of Volatile Phenols by α-CD, β-CD, and γ-CD in Model Wine*

Different CDs exhibited varying degrees of binding with the volatile phenols studied, i.e., guaiacol, 4-methylguaiacol, 4-ethylphenol, *o*-cresol, *m*-cresol, *p*-cresol, 4-ethylguaiacol, and eugenol (Table 3). In the current study, β-CD retained the highest proportion of volatile phenols, with the overall headspace concentration of volatile phenols reduced to 48.3% following the addition of 25 g/L β-CD. This was not unexpected, given β-CD is the most frequently reported inclusion complex host in other CD studies, because of its cavity size and hydrophobicity [19]. Guaiacol proved to be the most difficult compound to encapsulate within the CDs, with a 25 g/L addition of γ-CD achieving the best removal of guaiacol (being almost 30%). In contrast, 4-ethylphenol was most susceptible to CD complexation, and β-CD reduced the headspace concentration of 4-ethylphenol to just 23.1%. The ranking of volatile phenols by the extent to which their headspace concentration decreased following β-CD addition was: 4-ethylphenol > *p*-cresol > eugenol > *m*-cresol > 4-ethylguaiacol > *o*-cresol > 4-methylguaiacol > guaiacol. Differences in reactivity were attributed to the molecular structure of the volatile phenols. It has long been established that the hydrophobicity, molecular structure, and size of guest molecules are among the most influential factors in the formation of CD inclusion complexes [18,22]. Factors that influence binding between CDs and guest molecules have been previously studied [36]. In the current study, it was obvious that the molecular geometry and the polarity of chemical functional groups in the guest molecule played a major role in binding. The more highly retained 4-ethylphenol and *p*-cresol have the most "aligned" structure, with the non-polar alkyl groups attached in the *para* position of the benzene ring (Figure S2), whereas the less highly retained phenols, namely guaiacol and 4-methylguaiacol, have more polar methoxy groups at their *ortho* positions, which likely act to sterically hinder the molecule from entering into the β-CD cavity.

Table 3. Residual volatile phenol levels following addition of α-, β-, and γ-CDs to model wine, using four-phase HS-SPME GC-MS analysis.

		Guaiacol	4-Methylguaiacol	4-Ethylphenol	4-Ethylguaiacol	*o*-Cresol	*m*-Cresol	*p*-Cresol	Eugenol
Control		1.26 a ± 0.05	1.36 a ± 0.06	1.85 a ± 0.10	0.94 a ± 0.01	1.64 a ± 0.11	0.95 a ± 0.05	1.86 a ± 0.06	0.66 a ± 0.03
α-CD	5 g/L	1.14 abc ± 0.02 (90.3%)	1.19 ab ± 0.04 (87.9%)	1.51 ab ± 0.10 (81.6%)	0.82 b ± 0.01 (87.3%)	1.39 abc ± 0.06 (85.0%)	0.81 b ± 0.02 (85.1%)	0.92 b ± 0.03 (84.8%)	0.59 ab ± 0.03 (71.1%)
	25 g/L	1.17 ab ± 0.05 (92.9%)	1.20 ab ± 0.04 (88.2%)	1.42 b ± 0.11 (76.9%)	0.76 b ± 0.02 (80.4%)	1.41 ab ± 0.07 (86.2%)	0.75 bc ± 0.03 (79.3%)	0.85 bc ± 0.04 (78.3%)	0.47 c ± 0.01 (71.1%)
β-CD	5 g/L	1.01 cde ± 0.03 (80.2%)	1.04 bcd ± 0.04 (76.7%)	0.80 c ± 0.03 (43.1%)	0.68 c ± 0.02 (72.6%)	1.14 cd ± 0.03 (70.0%)	0.61 d ± 0.01 (63.9%)	0.61 d ± 0.01 (56.5%)	0.46 c ± 0.01 (70.0%)
	25 g/L	0.98 de ± 0.00 (77.5%)	0.95 cd ± 0.02 (69.8%)	0.43 d ± 0.02 (23.1%)	0.47 e ± 0.01 (50.0%)	0.92 d ± 0.01 (56.0%)	0.41 e ± 0.01 (42.6%)	0.34 e ± 0.01 (31.1%)	0.24 e ± 0.01 (36.9%)
γ-CD	5 g/L	1.07 bcd ± 0.01 (84.8%)	1.10 bc ± 0.01 (81.1%)	1.29 b ± 0.01 (69.8%)	0.79 b ± 0.00 (84.2%)	1.24 bc ± 0.01 (75.7%)	0.67 cd ± 0.01 (70.1%)	0.76 c ± 0.01 (70.0%)	0.56 b ± 0.01 (84.7%)
	25 g/L	0.89 e ± 0.02 (70.6%)	0.87 d ± 0.03 (64.1%)	0.77 cd ± 0.03 (41.5%)	0.57 d ± 0.01 (60.0%)	0.93 d ± 0.01 (57.2%)	0.49 e ± 0.01 (51.1%)	0.56 d ± 0.01 (51.5%)	0.35 d ± 0.01 (53.0%)

Values are means of three replicates ± standard error (and percentage of control). Values followed by different letters within columns are statistically significant (one-way ANOVA, p = 0.05).

3. Materials and Methods

3.1. Chemicals

Analytical grade volatile phenols (guaiacol, 4-methylguaiacol, 4-ethylguaiacol, 4-ethylphenol, *o*-cresol, *m*-cresol, *p*-cresol, and eugenol) and deuterated NMR solvents (d_6-ethanol, D_2O, and DCl) were purchased from Sigma-Aldrich (Castle Hill, NSW, Australia). Deuterium-labelled internal standards (d_3-guaiacol, d_3-4-methylguaiacol, and d_4-4-ethylphenol) were sourced from CDN Isotopes (Pointe-Claire, Quebec, Canada). Analytical grade ethanol, tartaric acid, and sodium hydroxide were purchased from Thermo Fisher Scientific (Waltham, MA, USA). Food grade (>98% purity) α-, β-, and γ-CDs were supplied by IMCD Group (Adelaide, SA, Australia). Model wine was prepared by dissolving tartaric acid (5 g/L) in aqueous ethanol (12% alcohol by volume) and adjusting the pH to 3.5 by dropwise addition of 1 M sodium hydroxide. Stock solutions of internal standards and volatile phenols were prepared volumetrically in absolute ethanol and stored at −20 °C, with working solutions prepared in model wine and stored at 4 °C.

3.2. Nuclear Magnetic Resonance Analysis

Complexation of volatile phenols by CDs was investigated by 2-dimensional nuclear magnetic resonance rotating frame Overhauser effect spectroscopy (^{1}H 2D ROESY). Samples were prepared by adding volatile phenols (10^{-3} mol/L) and CDs (10^{-2} mol/L) to deuterated model wine (i.e., 12% d_5-ethanol in D_2O, pD adjusted to 3.5 by dropwise addition of DCl). Spectra were recorded with an Agilent DD2 600 MHz spectrometer fitted with a cryoprobe (Agilent Technologies, Santa Clara, CA, USA) operating at 600 MHz with a delay time of 300 ms.

3.3. Sensory Analysis

Triangle tests [37] were performed to demonstrate the sensory impact of volatile phenol retention by β-CD. The panel comprised 38 postgraduate Wine Business students (8 male and 30 female, aged between 21 and 50 years) from the University of Adelaide. Model wines were presented in three-digit coded, covered XL5 wine glasses, using a balanced, randomized presentation order comprising all possible configurations, i.e., ABB, ABA, AAB, BAA, BAB, and BBA, where A denotes model wine spiked with volatile phenols and B denotes model wine spiked with volatile phenols and treated with β-CD (10 g/L). Panelists evaluated two brackets of wines: one representing smoke taint, comprising model wines spiked with guaiacol, 4-methylguaiacol, and *p*-cresol (1 mg/L each); and one representing *Brettanomyces* spoilage, comprising model wines spiked with 4-ethylphenol and 4-ethylguaiacol (1 mg/L each). Panelists smelled but did not taste wines, then identified the sample in each bracket that was considered to be different.

3.4. GC-MS Instrumental Analysis

Analysis of samples was performed with an Agilent GC-MS system (Santa Clara, CA, USA) comprising a 7890A gas chromatograph equipped with a Gerstel MPS autosampler (Mülheim, Germany) coupled to a 5975C mass selective detector. A DB-Wax column (60 m, 0.25 mm id, 0.25 μm film thickness, Agilent J&W, Folsom, CA, USA) was used for separation. The carrier gas was helium (BOC Gas, Adelaide, SA, Australia), at a constant flow of 1.5 mL/min. The inlet temperature was set at 240 °C and the oven temperature started at 40 °C for 1 min, increased to 200 °C at 5 °C/min, and was held at 200 °C for 5 min, before being increased to 250 °C at 10 °C/min and remaining at 250 °C for 10 min, giving a total run time of 52 min. The transfer line was set at 230 °C and positive ion electron impact spectra at 70 eV were recorded in the range *m/z* 25 to 215 for scan runs. For quantification of volatile phenols, mass spectra were recorded in selected ion monitoring (SIM) mode. The ions monitored in SIM mode were: *m/z* 109, *124* for guaiacol; *m/z* 109, *127* for d_3-guaiacol; *m/z* 123, *138* for 4-methylguaiacol; *m/z* 126, *141* for d_3-4-methylguaiacol; *m/z* 77, 90, *108* for *o*-cresol; *m/z* 122, 137, *152* for 4-ethylguaiacol; *m/z* 77, *107* for *p*-cresol; *m/z* 79, *108* for *m*-cresol; *m/z* 77, *122* for 4-ethylphenol; *m/z*

77, *126* for d_4-4-ethylphenol; and *m/z* 149, *164* for eugenol; with italicized ions used for quantitation. Volatile phenol concentrations are reported as relative peak areas (RPA), i.e., as the ratio of the peak area of the analyte (A_s) relative to the peak area of the isotopic standard (A_i).

3.5. HS-SPME GC-MS Analysis of Volatile Phenols in Model Wine Following CD Addition

An HS-SPME GC-MS method developed by other studies for determination of volatile phenols in wine (Castro-Mejías et al. 2003) was initially employed in the current study to determine the changes in volatile phenol levels following CD addition in model wine. Model wine was spiked with guaiacol, 4-methylguaiacol, or 4-ethylphenol at 0, 0.25, 0.5, 0.75, 1.0, 1.25, 1.5, 1.75, and 2 mg/L, and an aliquot of normalizing internal standard solution (containing 100 mg/L each of d_3-guaiacol, d_3-4-methylguaiacol, and d_4-4-ethylphenol) added, prior to SPME GC-MS analysis to develop calibration functions (Table S1). High linearity was observed over the working range, with correlation coefficients greater than 0.9995. A preliminary experiment using the above SPME GC-MS method involving the addition of α-CD, β-CD, or γ-CD (of 25 g/L) to model wine solutions containing guaiacol, 4-methylguaiacol, and 4-ethylphenol (1 mg/L each) suggested no significant binding of volatile phenols by the CDs; i.e., no significant differences were observed between the RPAs of compound-to-standard for volatile phenols with and without CD addition (Table 1). However, the absolute peak areas of analytes (and internal standards) were observed to be considerably smaller in samples with CD addition, e.g., the peak areas of 4-ethylphenol and d_4-4-ethylphenol reduced from approximately 100,000 to 20,000 abundance. Initially this was thought to reflect either variation in fiber performance or fiber degradation, but subsequent sensory and NMR analyses (described below) confirmed CD binding of volatile phenols, which led to the conclusion that CDs were also binding to the isotopically labelled standards and prompted the development of a novel HS-SPME method, involving introduction of the internal standard solution via an additional liquid phase, so as to prevent inclusion of standards by CDs. This was achieved by inserting a 2-mL glass ampoule (Gerresheimer Shuangfang Pharmaceutical Packaging, Zhenjiang, China) containing the internal standard solution into the 20-mL headspace vial (Sigma Aldrich, Castle Hill, NSW, Australia), as shown in Figure S3. A series of experiments (using a solution of methylene blue) were performed to ensure there was no mixing of samples in the SPME vial and the glass ampoule (or vice-versa), during sample preparation, or transfer, agitation, and extraction (data not shown). The influence of internal standard volume, agitation, incubation (temperature and duration), and the duration of sample extraction on the repeatability and accuracy of the novel SPME method were also evaluated, as method development and validation.

3.6. Method Development for the Four-Phase HS-SPME GC-MS Method

3.6.1. Influence of Agitation, Internal Standard Volume, and Pre-Analysis Equilibration Time

An aliquot of model wine (6 mL) containing 1 mg/L of guaiacol, 4-methylguaiacol, and 4-ethylphenol was transferred into headspace vials. This volume maximized the sample being analyzed, while ensuring the ampoule remained submerged. The inserted normalizing standard solution contained three isotopic standards (d_3-guaiacol, d_3-4-methylguaiacol, and d_4-4-ethylphenol) at 10 mg/L each. In preliminary benchtop experiments, both agitation and the volume of inserted liquid were found to change during the pre-analysis equilibration time (data not shown). Therefore, a multiple factorial design was adopted to optimize the extraction conditions (performed in triplicate). Four different volumes of internal standard were used (0.1, 0.5, 1.0, and 2.0 mL). Samples were analyzed over a 24-h period (at 3-hr intervals, in triplicate) to determine the optimal equilibration time. Agitation, when used, was set at 250 rpm. The autosampler incubation and extraction times were 10 and 15 min, respectively, and the extraction temperature was 35 °C.

3.6.2. Influence of Extraction Time, Extraction Temperature, and Internal Standard Concentration

Using the optimized parameters identified above, several additional parameters were evaluated. Extraction temperatures of 35, 50, 65, and 80 °C, extraction times of 15, 30, 45, and 60 min, and internal standard concentrations of 5, 10, 20, 30, 40, and 50 mg/L were evaluated; with all samples prepared in triplicate.

3.7. Method Performance for the Four-Phase HS-SPME GC-MS Method

The optimized SPME method comprised the following conditions: an ampoule tube containing 0.5 mL of model wine spiked with 10 mg/L of internal standard solution was inserted into 6 mL of sample (i.e., model wine spiked with volatile phenols) in a 20-mL headspace sampling vial. Equilibrium in the headspace vial was achieved via 15-min incubation at 35 °C, before 15-min extraction without agitation. To validate the method, calibration curves were generated for guaiacol, 4-methylguaiacol, 4-ethylguaiacol, 4-ethylphenol, *o*-cresol, *m*-cresol, *p*-cresol, and eugenol, spiked at 0, 0.25, 0.5, 0.75, 1.0, 1.25, 1.5, 1.75, and 2.0 mg/L. Guaiacol, 4-methylguaiacol, and 4-ethylphenol were quantified against their isotopically labelled equivalents, whereas 4-ethylguaiacol, *o*-cresol, *m*-cresol, *p*-cresol, and eugenol were quantified against d_3-guaiacol. The linear range of detection was subsequently validated at volatile phenol concentrations up to 50 mg/L. Samples spiked with 0.25, 0.5, 1.0, 1.25, 1.75, or 2.0 mg/L of guaiacol, 4-methylguaiacol, and 4-ethylphenol were prepared in triplicate and analyzed to evaluate method repeatability; with preparation and analysis of a subset of samples (i.e., samples spiked at 0.5, 1.25, and 1.75 mg/L) repeated after 1 month. All validation samples were analyzed in triplicate.

3.8. Retention of Volatile Phenols in Model Wine by Cyclodextrins

A model wine solution comprising 1 mg/L of guaiacol, 4-methylguaiacol, 4-ethylguaiacol, 4-ethylphenol, *o*-cresol, *m*-cresol, *p*-cresol, and eugenol was prepared. Aliquots (6 mL) were placed in 20-mL SPME headspace vials, to which 25 g/L of α-CD, β-CD, or γ-CD were added. Samples were then heated to 35 °C in an incubator (Ratek, Boronia, VIC, Australia) with agitation (200 rpm) for 20 min, after which samples were cooled to ambient temperature and analyzed by GC-MS. Samples were prepared in triplicate. Control samples (i.e., samples without the addition of CD) were also prepared in triplicate. The residual volatile phenol levels were determined using the optimized four-phase SPME GC-MS method. Semi-quantification based on standard addition was used to calculate the percentage difference between the RPA of residual volatile phenols following CD addition, with those of control samples.

3.9. Data Analysis

Data are presented as mean values of three replicates ± standard error. One-way ANOVA was conducted to determine the differences between sample means, with a T-test at $p = 0.05$, using XLSTAT software (version 2015.3, Addinsoft, Paris, France).

4. Conclusions

The newly developed four-phase HS-SPME GC-MS method overcame the difficulties associated with CD encapsulation of internal standards spiked directly into samples. Although this method does not completely prevent interactions between isotopically labelled standards and dissolved CDs, it mitigates interactions by introducing the standard via a separate liquid phase. Modification and validation therefore need to be undertaken when adapting this method for the analysis of other volatile compounds and in more complicated substrates (e.g., wine, rather than model wine). The improvements offered by this method nevertheless enabled complexation between CDs and volatile phenols to be studied. CDs were found to form inclusion complexes with volatile phenols in model wine, resulting in reductions in the perceived intensity of off-odors. Importantly, this method can be adapted for quantitative analysis of other systems in which a substrate component might similarly scavenge internal standards.

Supplementary Materials: The following are available online at http://www.mdpi.com/1420-3049/24/19/3432/s1, Table S1: Calibration curve using conventional HS-SPME GC-MS method. Figure S1: Structures of α-CD, β-CD, and γ-CD. Figure S2: Encapsulation of 4-ethylphenol within β-CD. Figure S3: Diagram of headspace vial containing model wine sample, with different volumes of internal standard in the glass ampoule (as indicated by shading).

Author Contributions: All authors were involved in conceptualization; methodology, C.D.; validation, C.D.; formal analysis, C.D.; investigation, C.D.; writing—original draft preparation, C.D. and K.L.W.; writing—review and editing, V.J. and D.K.T.; supervision, K.L.W., V.J. and D.K.T.; funding acquisition, K.L.W., V.J. and D.K.T.

Funding: This research was conducted by the Australian Research Council Training Centre for Innovative Wine Production (www.adelaide.edu.au/tc-iwp), which is funded as part of the ARC's Industrial Transformation Research Program (Project No. ICI30100005), with support from Wine Australia and industry partners.

Acknowledgments: The authors gratefully acknowledge: CSIRO's Paul Boss and Emily Nicholson for technical assistance with GC-MS analysis; and the students who participated in sensory analysis.

Conflicts of Interest: The authors declare no conflict of interest. The funders had no role in the design of the study, i.e., in the collection, analysis or interpretation of data, in the writing of the manuscript, or in the decision to publish the results.

References

1. Maga, J.A. The contribution of wood to the flavour of alcoholic beverages. *Food Rev. Int.* **1989**, *5*, 39–99. [CrossRef]
2. Boidron, J.N.; Chatonnet, P.; Pons, M. Influence du bois sur certaines substances odorantes des vins. *Conn. Vigne Vin* **1988**, *22*, 275–294. [CrossRef]
3. Prida, A.; Chatonnet, P. Impact of oak-derived compounds on the olfactory perception of barrel-aged wines. *Am. J. Enol. Vitic.* **2010**, *61*, 408–413.
4. Chatonnet, P.; Dubourdieu, D.; Boidron, J.N. The influence of *Brettanomyces/Dekkera* sp. yeasts and lactic acid bacteria on the ethylphenol content of red wines. *Am. J. Enol. Vitic.* **1995**, *46*, 463–468.
5. Kennison, K.R.; Wilkinson, K.L.; Williams, H.G.; Smith, J.H.; Gibberd, M.R. Smoke-derived taint in wine: Effect of postharvest smoke exposure of grapes on the chemical composition and sensory characteristics of wine. *J. Agric. Food Chem.* **2007**, *55*, 10897–10901. [CrossRef]
6. Kennison, K.R.; Gibberd, M.R.; Pollnitz, A.P.; Wilkinson, K.L. Smoke-derived taint in wine: The release of smoke-derived volatile phenols during fermentation of Merlot juice following grapevine exposure to smoke. *J. Agric. Food Chem.* **2008**, *56*, 7379–7383. [CrossRef] [PubMed]
7. Kennison, K.R.; Wilkinson, K.L.; Pollnitz, A.P.; Williams, H.G.; Gibberd, M.R. Effect of timing and duration of grapevine exposure to smoke on the composition and sensory properties of wine. *Aust. J. Grape Wine Res.* **2009**, *15*, 228–237. [CrossRef]
8. Chassagne, D.; Guilloux-Benatier, M.; Alexandre, H.; Voilley, A. Sorption of wine volatile phenols by yeast lees. *Food Chem.* **2005**, *91*, 39–44. [CrossRef]
9. Pradelles, R.; Alexandre, H.; Ortiz-Julien, A.; Chassagne, D. Effects of yeast cell-wall characteristics on 4-ethylphenol sorption capacity in model wine. *J. Agric. Food Chem.* **2008**, *56*, 11854–11861. [CrossRef]
10. Fudge, A.L.; Schiettecatte, M.; Ristic, R.; Hayasaka, Y.; Wilkinson, K.L. Amelioration of smoke taint in wine by treatment with commercial fining agents. *Aust. J. Grape Wine Res.* **2012**, *18*, 302–307. [CrossRef]
11. Lisanti, M.T.; Gambuti, A.; Genovese, A.; Piombino, P.; Moio, L. Treatment by fining agents of red wine affected by phenolic off-odour. *Eur. Food Res. Technol.* **2017**, *243*, 501–510. [CrossRef]
12. Ugarte, P.; Agosin, E.; Bordeu, E.; Villalobos, J.I. Reduction of 4-ethylphenol and 4-ethylguaiacol concentration in red wines using reverse osmosis and adsorption. *Am. J. Enol. Vitic.* **2005**, *56*, 30–36.
13. Fudge, A.L.; Ristic, R.; Wollan, D.; Wilkinson, K.L. Amelioration of smoke taint in wine by reverse osmosis and solid phase adsorption. *Aust. J. Grape Wine Res.* **2011**, *17*, S41–S48. [CrossRef]
14. Larcher, R.; Puecher, C.; Rohregger, S.; Malacarne, M.; Nicolini, G. 4-Ethylphenol and 4-ethylguaiacol depletion in wine using esterified cellulose. *Food Chem.* **2012**, *132*, 2126–2130. [CrossRef]

15. Carrasco-Sánchez, V.; John, A.; Marican, A.; Santos, L.S.; Laurie, V.F. Removal of 4-ethylphenol and 4-ethylguaiacol with polyaniline-based compounds in wine-like model solutions and red wine. *Molecules* **2015**, *20*, 14312–14325. [CrossRef] [PubMed]
16. Garde-Cerdán, T.; Zalacain, A.; Lorenzo, C.; Alonso, J.L.; Salinas, M.R. Molecularly imprinted polymer-assisted simple clean-up of 2, 4, 6-trichloroanisole and ethylphenols from aged red wines. *Am. J. Enol. Vitic.* **2008**, *59*, 396–400.
17. Teixeira, R.; Dopico-García, S.; Andrade, P.B.; Valentão, P.; López-Vilariño, J.M.; González-Rodríguez, V.; Cela-Pérez, C.; Silva, L.R. Volatile phenols depletion in red wine using molecular imprinted polymers. *J. Food Sci. Technol.* **2015**, *52*, 7735–7746. [CrossRef]
18. Szejtli, J. Introduction and general overview of cyclodextrin chemistry. *Chem. Rev.* **1998**, *98*, 1743–1754. [CrossRef]
19. Kant, A.; Linforth, R.S.T.; Hort, J.; Taylor, A.J. Effect of β-cyclodextrin on aroma release and flavor perception. *J. Agric. Food Chem.* **2004**, *52*, 2028–2035. [CrossRef]
20. Buschmann, H.-J.; Schollmeyer, E. Applications of cyclodextrins in cosmetic products: A review. *J. Cosmetic Sci.* **2002**, *53*, 185–191.
21. Challa, R.; Ahuja, A.; Ali, J.; Khar, R.K. Cyclodextrins in drug delivery: An updated review. *AAPS Pharm. Sci. Tech.* **2005**, *6*, E329–E357. [CrossRef] [PubMed]
22. Astray, G.; Gonzalez-Barreiro, C.; Mejuto, J.C.; Rial-Otero, R.; Simal-Gándara, J. A review on the use of cyclodextrins in foods. *Food Hydrocoll.* **2009**, *23*, 1631–1640. [CrossRef]
23. Marques, H.M.C. A review on cyclodextrin encapsulation of essential oils and volatiles. *Flavour Fragr. J.* **2010**, *25*, 313–326. [CrossRef]
24. Botelho, G.; Valiau, C.; Moreira da Silva, A. Effect of cyclodextrins on off-odours removal of red wine: An innovative approach. *Ciência e Técnica Vitivinícola* **2011**, *26*, 63–68.
25. Ratnasooriya, C.C.; Rupasinghe, H.P.V. Extraction of phenolic compounds from grapes and their pomace using β-cyclodextrin. *Food Chem.* **2012**, *134*, 625–631. [CrossRef] [PubMed]
26. Whiton, R.S.; Zoecklein, B.W. Optimization of headspace solid-phase microextraction for analysis of wine aroma compounds. *Am. J. Enol. Vitic.* **2000**, *51*, 379–382.
27. Rocha, S.; Ramalheira, V.; Barros, A.; Delgadillo, I.; Coimbra, M.A. Headspace solid phase microextraction (SPME) analysis of flavor compounds in wines. Effect of the matrix volatile composition in the relative response factors in a wine model. *J. Agric. Food Chem.* **2001**, *49*, 5142–5151. [CrossRef]
28. Castro-Mejías, R.; Natera-Marín, R.; Valme García-Moreno, M.; Barroso, C.G. Optimisation of headspace solid-phase microextraction for the analysis of volatile phenols in wine. *J. Chromatogr. A* **2003**, *995*, 11–20. [CrossRef]
29. Tsoutsi, C.; Konstantinou, I.; Hela, D.; Albanis, T. Screening method for organophosphorus insecticides and their metabolites in olive oil samples based on headspace solid-phase microextraction coupled with gas chromatography. *Anal. Chim. Acta* **2006**, *573*, 216–222. [CrossRef]
30. Pawliszyn, J.; Yang, M.J.; Orton, M.L. Quantitative determination of caffeine in beverages using a combined SPME-GC/MS method. *J. Chem. Edu.* **1997**, *74*, 1130. [CrossRef]
31. Pawliszyn, J. Theory of solid-phase microextraction. *J. Chromatogr. Sci.* **2000**, *38*, 270–278. [CrossRef] [PubMed]
32. Sander, R. Compilation of Henry's law constants (version 4.0) for water as solvent. *Atmos. Chem. Phys.* **2015**, *15*, 4399–4981. [CrossRef]
33. Ma, S.; Turino, G.M.; Hayashi, T.; Yanuma, H.; Usuki, T.; Lin, Y.Y. Stable deuterium internal standard for the isotope-dilution LC–MS/MS analysis of elastin degradation. *Anal. Biochem.* **2013**, *440*, 158–165. [CrossRef] [PubMed]
34. Wieland, F.; Neff, A.; Gloess, A.N.; Poisson, L.; Atlan, S.; Larrain, D.; Prêtre, D.; Blank, I.; Yeretzian, C. Temperature dependence of Henry's law constants: An automated, high-throughput gas stripping cell design coupled to PTR-ToF-MS. *Int. J. Mass Spectrom.* **2015**, *387*, 69–77. [CrossRef]
35. Liang, H.R.; Foltz, R.L.; Meng, M.; Bennett, P. Ionization enhancement in atmospheric pressure chemical ionization and suppression in electrospray ionization between target drugs and stable-isotope-labeled internal standards in quantitative liquid chromatography/tandem mass spectrometry. *Rapid Commun. Mass Spectrom.* **2003**, *17*, 2815–2821. [CrossRef] [PubMed]

36. Astray, G.; Mejuto, J.C.; Morales, J.; Rial-Otero, R.; Simal-Gándara, J. Factors controlling flavors binding constants to cyclodextrins and their applications in foods. *Food Res. Int.* **2010**, *43*, 1212–1218. [CrossRef]
37. Meilgaard, M.C.; Carr, B.T.; Civille, G.V. *Sensory Evaluation Techniques*, 4th ed.; CRC Press: Boca Raton, FL, USA, 2007.

Sample Availability: Samples of the compounds are not available from the authors.

molecules

Article

Assessment of the Bioactive Compounds in White and Red Wines Enriched with a *Primula veris* L.

Maria Tarapatskyy [1,*], Ireneusz Kapusta [2], Aleksandra Gumienna [1] and Czesław Puchalski [1]

[1] Department of Bioenergetics and Food Analysis, Faculty of Biology and Agriculture, University of Rzeszów, 35-601 Rzeszów, Poland; gumienna.ola@wp.pl (A.G.); cpuchal@ur.edu.pl (C.P.)
[2] Department of Food Technology and Human Nutrition, Faculty of Biology and Agriculture, University of Rzeszów, 35-601 Rzeszów, Poland; ikapusta@ur.edu.pl
* Correspondence: mczernicka@ur.edu.pl

Academic Editors: Rosa Maria de Sá Perestrelo and José Sousa Câmara
Received: 1 October 2019; Accepted: 8 November 2019; Published: 11 November 2019

Abstract: The aim of this paper was to analyze selected physicochemical properties and the pro-health potential of wines produced in southeastern Poland, in the Subcarpathian region, and commercial Carlo Rossi wines enhanced with cowslip (*Primula veris* L.). This study used ultra-performance reverse-phase liquid chromatography (UPLC)-PDA-MS/MS to perform most of the analysis, including the polyphenolic compounds and saponin content in wines enriched by *Primula veris* L. The initial anthocyanin content in Subcarpathian (Regional) red wine samples increased four times to the level of 1956.85 mg/L after a 10% addition of *Primula veris* L. flowers. For white wines, a five-fold increase in flavonol content was found in Subcarpathian (Regional) and wine samples, and an almost 25-fold increase in flavonol content was found in Carlo Rossi (Commercial) wine samples at the lowest (2.5%) *Primula veris* L. flower addition. Qualitative analysis of the regional white wines with a 10% *Primula veris* L. flower enhancement demonstrated the highest kaempferol content (197.75 mg/L) and a high quercetin content (31.35 mg/L). Thanks to wine enrichment in triterpenoid saponins and in polyphenolic compounds from *Primula veris* L. flowers, which are effectively extracted to wine under mild conditions, both white and red wines can constitute a highly pro-health component of diets, which is valuable in preventing chronic heart failure.

Keywords: wine; *Primula veris* L.; cowslip; health potential; flavonoids; anthocyanins; triterpenoid saponins; antioxidant activity

1. Introduction

Wine is an alcoholic beverage produced as a result of alcoholic fermentation of fresh *Vitis vinifera* grapevine fruits (fragmented or not) or must. There are multiple wine classifications, e.g., by color, strength, or total sugar content. The decisive factor affecting the flavor, aroma, and composition of wine is the variety of the grapevine from which the wine is made. Other factors are natural environment characteristics, such as soil, climate, method of cultivation, vinification process technology, and many others [1].

The main components of wine are water (approximately 86%) and ethyl alcohol (12% on average), accompanied by much smaller amounts of glycerol, sugars, organic acids, mineral compounds, vitamins, tannins, and biologically active compounds, which are also responsible for wine's pro-health properties, such as its anti-oxidant, anti-carcinogenic, anti-inflammatory, immunomodulating, anti-virus and anti-bacterial effects [2,3]. It has also been demonstrated that moderate wine intake greatly reduces the risk of cardiovascular diseases, although due to the alcohol content, international guidelines limit wine consumption to the level of 150–250 mL (15–30 g alcohol) [2].

The increased interest in wine production in Poland is related to climate changes (a rising trend in temperature and shorter winters have been observed), the emergence of more resistant grapevine varieties, greater knowledge concerning grapevine cultivation, increased consumer awareness of wine's pro-health properties, and farmers seeking new, profitable crops [4,5].

Grapevine cultivation in Poland is not the easiest task as the climate differs significantly from the conditions that are characteristic of typical viticultural regions. Despite this, winemaking is a growing industry in Poland, where the tradition reaches as far as the Middle Ages [4]. The climate in Poland is characterized by significant daily and seasonal fluctuations in temperature, with a risk of frost and hailstorms during the spring season [5]. The European Union has classified Poland as viticultural Zone A, i.e., the coldest one, together with Germany (except Baden), the Czech Republic (excluding Moravia), Belgium, and the United Kingdom [6], which means that wines from Poland can be sold on the European market.

The Subcarpathian region, located in the southeastern part of the country (number 6 in Figure 1) is considered exceptionally important for winemaking in Poland. At present, close to 150 vineyards are located there, and most of them environmentally friendly, whose total area is estimated at over 100 hectares [7,8]. Thanks to specific terrain features (foothills with gentle slopes, clay soils, poorly industrialized) and microclimate (hot summers and sunny autumns), the region has become a perfect location for the development of winemaking, as well as winemaking tourism [8].

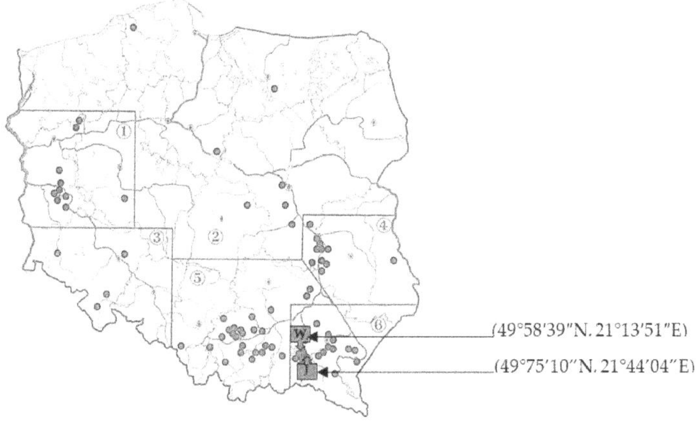

Figure 1. Wine Regions in Poland with some of the major vineyards highlighted (points). 1—Zielona Góra; 2—Central and Northern Poland; 3—Lower Silesia; 4—Lesser Poland Gorge of the Vistula (with Lublin Voivodeship); 5—Lesser Poland (with Śląskie and Świętokrzyskie Voivodships); 6—Podkarpacie.

In Poland, grapevine varieties that are hybrids of *Vitis vinifera* and species originating from North America (*Vitis labrusca* and *V. aestivalis*, so-called American hybrids; or *V. rupestris* and *V. lincekumii*, so-called French-American hybrids), or Asia (e.g., *V. amurensis*, the Amur grape) are the most commonly cultivated. These varieties are characterized not only by resistance to low temperatures but also to diseases, as well as high and regular yields [9]. The wines they produce exhibit a specific flavor and aroma, which stem from the slightly different chemical composition than wines produced from *V. vinifera*. The colder climate and shorter summers lead to grape fruits usually having a lower sugar content and higher acidity. Cold-climate wines are viewed as subtler and more "refined", and the higher acidity imbues them with freshness [9]. For this reason, winemaking in the Subcarpathian region continues to grow, and the local wines are unique products that enjoy growing recognition on the market. As winemaking technology in Poland has evolved across decades, alternative uses for wine have been sought, particularly due to its pro-health properties. One such procedure has been enhancing wine with cowslip (*Primula veris* L.), which was supposed to enrich the wine with substances of high biological activity.

Polyphenolic compounds are commonly known as plant secondary metabolites that hold an aromatic ring bearing at least one hydroxyl group. These phytochemical substances are presented in nutrients and herbal medicines; many studies have reported on both flavonoids and other phenolic components due to their effective antioxidant, anticancer, anti-inflammation, and antibacterial properties; their function as cardioprotective agents; their immune system-promoting and skin protection from ultraviolet (UV) radiation factors, and their interesting possibilities for pharmaceutical and medical applications [10–12].

Over the last years, polyphenol-rich foods and polyphenols have received great attention due to their potential beneficial effects toward human health. Contained not only in fruits and vegetables, but also in whole grains, nuts, olive oil, and beverages such as coffee and tea, they are characteristic components of healthy dietary patterns. Recent evidence has proposed that a higher dietary intake of polyphenols may be inversely associated with overall and cardiovascular disease-related mortality, certain cancers, cardiovascular diseases, anthropometric measures, and mood disorders [13]. These features make polyphenols a potentially interesting material for the development of functional foods or possible therapy for the prevention of some diseases. The health effects of polyphenols depend on both their respective intakes and their bioavailability, which can vary greatly. Numerous genetic, environmental, and technologic factors may affect the polyphenol concentrations in food, some of which can be controlled to optimize the polyphenol content of foods. One of the possibilities for the increase of phenolic compounds in food may be the fruits and vegetables cultivar selection, as cultivars differ greatly in their phenolics content. The strong potential may also be the processing technology polyphenol fortification of food products, which must be taken into consideration.

Cowslip (*Primula veris* L.) is a plant species of the Primulaceae family. Until 2014, this plant was protected in Poland. According to the literature data, it is an herbal material that is rich in saponins (approximately 60%), including primulic acid 1 (PA 1), as well as numerous flavonoid compounds and flavonols, i.e., rutin, catechin, kaempferol, and luteolin [14,15]. In folk medicine, cowslip decoctions had been used as a sedative in migraine, insomnia, nervous stress, and menstruation ailment treatment. The diuretic properties of this herb had also been known, which led to its use in organism purging and detoxification treatments. Cowslip-enhanced wine, in turn, has been recommended for regulating blood circulation and in post-stroke and post-hemorrhage recovery. It is suspected that these properties may also be the result of the coumarin derivatives present in the herb. Cowslip's bioactive substances have found use in pharmaceutical preparations, e.g., in mucoactive syrups and in preparations used in respiratory system diseases [16,17]. As Poland does not belong to leading wine producers, the production of specialist wines enhanced with pro-health additions in the form of green plants rich in bioactive agents is worthy of consideration. Therefore, the aim of this paper was to analyze selected physicochemical properties and the pro-health potential of wines produced in southeastern Poland, in the Subcarpathian region, and commercial Carlo Rossi wines enhanced with cowslip (*Primula veris* L.). Furthermore, polyphenolic compounds and saponins in herb-enhanced wine samples were identified. The tests were intended to determine the effectiveness of pro-health substance extraction in different wines and at different cowslip (*Primula veris* L.) content levels. Additionally, ethanol blends of different strengths were analyzed to study the effects of alcohol content on pro-health substance extraction from cowslip. An analysis of pro-health substance content may aid in determining the cardioprotective effects of white and red wines enhanced with *Primula veris* L., whose consumption may constitute an effective practice in preventing chronic heart failure.

2. Results

Based on MS chromatography analysis, profiles of compounds belonging to anthocyanins (28), flavonols (22), and saponins (3) have been identified in the red wine samples, as well as 22 flavonols and three saponins in white wine samples. All the compounds identified are shown in Tables 1 and 2, while Table 3 presents the concentrations of individual compounds, the polyphenolic compounds of the anthocyanin group, as well as their total content in regional and commercial red wines enhanced with *Primula veris* L.

Anthocyanins constituted the dominant class of polyphenolic compounds in the test red wines, and their content ranged from 492.25 mg/L in Polish regional wines to 682.00 mg/L in popular Californian wines. Despite the higher total anthocyanin content in the test commercial wines, their profile differed from the regional wines in the lower content of delphinidin, cyanidin, and petunidin derivatives. Furthermore, a major drop in anthocyanin content was observed in all *Primula veris* L. enhanced commercial wine samples after 14 days of extraction, and the color of commercial wine blends enhanced with *Primula veris* L. flowers became brighter. The greatest drop in anthocyanin content, approximately six-fold, was found in samples with the highest addition of *Primula veris* L. flowers (10%), compared to the commercial wines without the flower addition. On the other hand, the lowest, four-fold reduction in anthocyanin content was noted for samples with a 2.5% addition of *Primula veris* L. The reason for the major reduction in anthocyanin content in the enhanced commercial wine samples was most likely their dilution with water contained in fresh *Primula veris* L. flowers, the emergence of conditions favorable for the deactivation/decomposition of pigments in red wines, and their high instability during wine sample storage at room temperature during the 14-day extraction. In this respect, regional wines demonstrated not only a higher stability, but also a high ability to extract anthocyanins from *Primula veris* L. The initial anthocyanin content in regional wine samples increased four times to the level of 1956.85 mg/L for a 10% addition of *Primula veris* L. flowers. In the case of a 2.5% addition of these flowers, the anthocyanin content in regional wine samples increased almost twice, compared to non-enhanced wines.

A high content of polyphenolic compounds of the flavonol group was found in both the regional and commercial red wine samples. The mean content of these substances is shown in Table 4.

The regional wines were characterized by a flavonol content approximately 62% lower than the commercial wines, in which the content of these substances was on the level of 104.55 mg/L. At the lowest *Primula veris* L. flower addition (2.5%), the flavonols content in the regional wines increased almost 10-fold, while at the 10% *Primula veris* L. addition, the flavonols content was 1883.60 mg/L, which is almost 30 times higher than in wine samples without the addition of *Primula veris* L. flowers. In the commercial red wines, the flower content led to an increase in total flavonol content to the level of 366 mg/L, and for the 10% addition of *Primula veris* L. flowers, the content of these substances increased 14 times. An increase in kaempferol and quercetin was observed in the same samples, to the levels of 100.50 mg/L and 38.05 mg/L, respectively, while in the regional wines with a 10% *Primula veris* L. enhancement, kaempferol content was almost four times lower, and quercetin almost three times lower than in similar samples of enhanced commercial wines. The regional wines exhibited superior extraction abilities in relation to polyphenolic compounds from *Primula veris* L. flowers, but a lower content of health-crucial anti-oxidants and their derivatives. An analysis of the polyphenolic compound profile in the white wines, shown in Table 5, confirmed the excellent extraction abilities of both regional and commercial wines.

For white wines, a five-fold increase in flavonol content was found in regional and wine samples, and an almost 25-fold increase in commercial wine samples at the lowest *Primula veris* L. flower addition of 2.5%. When comparing the results of the polyphenol profile analysis, the white wines can be considered more effective extractants than the test red wines. In the case of white wines, the mean flavonol content ranged from 92.50 mg/L in the regional wines to 64.35 mg/L in the commercial wines, and at the 10% addition of *Primula veris* L. flowers, the flavonol content was 2356.80 mg/L and 2404.30 mg/L, respectively, which exceeded the highest content of these substances in the red regional wines by approximately 27%. Qualitative analysis of the regional white wines with a 10% *Primula veris* L. flower enhancement demonstrated the highest kaempferol content (197.75 mg/L), and a high quercetin content (31.35 mg/L), while for the white commercial wines enhanced with the same amount of flowers, the content of these substances was eight times lower for kaempferol and 2.6 times lower for quercetin.

Table 1. Individual anthocyanins identified by ultra-performance reverse-phase liquid chromatography (UPLC)-PDA-MS/MS in red wine.

No	Compound	RT (min)	$[M-H]^+$ (*m/z*)	Fragment Ions (*m/z*)	Absorbance Maxima (nm)
1	Delphinidin 3-*O*-glucoside-5-*O*-glucoside	2.04	627	465, 303	277, 525
2	Cyanidin 3-*O*-glucoside-5-*O*-glucoside	2.19	611	449, 287	280, 516
3	Delphinidin 3-*O*-glucoside	2.38	465	303	280, 523
4	Petunidin 3-*O*-glucoside-5-*O*-glucoside	2.53	641	479, 317	277, 531
5	Peonidin 3-*O*-glucoside-5-*O*-glucoside	2.67	625	463, 301	278, 513
6	Malvidin 3-*O*-glucoside-5-*O*-glucoside	2.72	655	493, 331	275, 524
7	Cyanidin 3-*O*-glucoside	2.74	449	287	279, 515
8	Petunidin 3-*O*-glucoside	2.92	479	317	277, 526
9	Peonidin 3-*O*-glucoside	3.31	463	301	279, 515
10	Malvidin 3-*O*-glucoside	3.43	493	331	278, 530
11	Delphinidin 3-*O*-(6″-*O*-acetyl)-glucoside	3.53	507	465, 303	280, 528
12	Cyanidin 3-*O*-(6″-*O*-acetyl)-glucoside	3.95	491	449, 287	283, 522
13	Petunidin 3-*O*-(6″-*O*-acetyl)-glucoside	4.1	521	317	280, 530
14	Petunidin 3-*O*-(6″-*O*-acetyl)-glucoside-5-*O*-glucoside	4.28	787	625, 479, 317	280, 530
15	Delphinidin 3-*O*-(6″-*O*-coumaryl)-glucoside	4.47	611	303	279, 530
16	Malvidin 3-*O*-(6″-*O*-acetyl)-glucoside	4.62	535	331	280, 521
17	Malvidin 3-*O*-(6″-*O*-coumaryl)-glucoside-5-*O*-glucoside	4.67	801	639, 493, 331	280, 530
18	Peonidin 3-*O*-(6″-*O*-coumaryl)-glucoside-5-*O*-glucoside	4.68	771	609, 463, 301	279, 523
19	Peonidin 3-*O*-(6″-*O*-acetyl)-glucoside	4.85	505	463, 301	277, 535
20	Cyanidin 3-*O*-(6″-*O*-coumaryl)-glucoside	4.93	595	287	283, 522
21	Petunidin 3-*O*-(6″-*O*-coumaryl)-glucoside	4.98	625	317	280, 531
22	Delphinidin 3-*O*-(6″-caffeoyl)-glucoside	5.35	627	465, 303	280, 528
23	Peonidin 3-*O*-(6″-*O*-coumaryl)-glucoside	5.39	609	301	279, 523
24	Malvidin 3-*O*-(6″-*O*-coumaryl)-glucoside	5.44	639	331	280, 521

Table 2. Individual flavonols and saponins identified by (UPLC)-PDA-MS/MS in wine and ethanol.

No	Compound	Rt (min.)	[M − H]$^+$ (*m/z*)	Fragment Ions (*m/z*)	Absorbance Maxima (nm)
1	Quercetin 3-*O*-(6″-*O*-rhamnosyl)-glucoside 7-*O*-rhamnoside	4.13	755	609, 301	254, 353
2	Metyl-miricetin-3-*O*-rutinoside	4.28	639	493, 331, 317	276, 339
3	Metyl-miricetin-*O*-glucoside	4.33	493	331, 317	276, 338
4	Kaempferol 3-O-rutinoside 7-*O*-rhamnoside	4.50	739	593, 285	264, 347
5	Metyl-quercetin 3-*O*-(6″-*O*-rhamnosyl)-glucoside 7-*O*-rhamnoside	4.58	769	755, 609, 301	253, 354
6	Quercetin 3-*O*-rutinoside	4.65	609	301	255, 354
7	Quercetin 3-*O*-glucoside	4.79	463	301	255, 355
8	Kaempferol 3-*O*-rutinoside	4.84	593	285	262, 352
9	Kaempferol 3-*O*-rhamnosyl-glucoside isomer I	4.96	593	285	264, 348
10	Kaempferol 3-*O*-rhamnosyl-glucoside isomer II	5.15	593	285	264, 347
11	Metyl-quercetin 3-*O*-rutinoside	5.21	623	609, 301	253, 353
12	Metyl-quercetin 3-*O*-rhamnosyl-glucoside	5.28	623	609, 301	255, 354
13	Undefined kaempferol derivative	5.48	507	493, 285	269, 355
14	Quercetin 3-*O*-glucuronide	5.52	477	301	255, 354
15	Kaempferol 3-*O*-pentosyl-pentoside	5.87	549	285	269, 355
16	Kaempferol	6.94	285	-	264, 367
17	Quercetin	7.02	301	-	255, 355
18	Isorhamnetin	8.35	315	-	255, 355
1	Primula saponin II	4.19	1236	924, 465	
2	Primula saponin I	4.31	1104	924, 465, 447, 246	
3	Priverosaponin B 22-acetate	4.53	1162	982, 465	

Table 3. The content of anthocyanins in red wines enriched with *Primula veris* L. (mg/L).

Red Wine	**Regional**				**Commercial**			
***Primula veris* L. (%)**	**0**	**2.5**	**5**	**10**	**0**	**2.5**	**5**	**10**
Anthocyanins								
1	0.11 ± 0.05	22.15 ± 0.50	23.45 ± 0.35	30.70 ± 0.85	tr	tr	tr	tr
2	0.06 ± 0.11	4.95 ± 0.05	5.65 ± 0.15	5.72 ± 0.25	tr	tr	tr	tr
3	1.44 ± 0.30	6.85 ± 0.00	7.72 ± 0.05	9.65 ± 0.61	tr	tr	tr	tr
4	0.43 ± 0.35	30.95 ± 0.30	35.61 ± 1.30	44.45 ± 0.92	tr	tr	tr	tr
5	0.10 ± 0.00	21.50 ± 0.70	28.15 ± 0.35	31.71 ± 3.30	15.02 ± 0.25	14.45 ± 0.15	tr	tr
6	6.22 ± 0.20	19.31 ± 0.15	42.55 ± 1.87	38.22 ± 15.95	14.45 ± 0.20	14.02 ± 0.35	13.42 ± 1.11	tr
7	3.55 ± 0.65	5.65 ± 0.20	6.25 ± 0.25	7.50 ± 0.30	4.53 ± 0.15	tr	tr	tr
8	6.16 ± 0.00	7.77 ± 0.55	16.55 ± 0.11	18.81 ± 1.65	6.61 ± 0.20	tr	5.30 ± 0.85	tr
9	0.52 ± 2.05	18.11 ± 0.20	20.75 ± 1.35	25.62 ± 0.65	15.70 ± 0.20	15.01 ± 0.40	39.7 ± 2.01	tr
10	0.52 ± 0.10	224.65 ± 8.31	306.70 ± 14.95	343.90 ± 24.31	52.04 ± 0.20	3.65 ± 0.05	4.42 ± 0.52	9.4 ± 0.25
11	2.61 ± 0.70	24.50 ± 0.30	27.22 ± 0.10	33.35 ± 0.92	23.12 ± 0.82	tr	tr	tr
12	1.20 ± 0.01	7.15 ± 0.55	8.75 ± 0.31	9.85 ± 0.55	6.92 ± 0.05	5.85 ± 0.05	5.21 ± 0.95	tr
13	0.81 ± 0.15	tr	tr	tr	4.85 ± 0.01	tr	tr	tr
14	6.45 ± 0.05	10.75 ± 0.10	12.75 ± 0.35	14.65 ± 2.55	6.65 ± 0.02	6.25 ± 0.12	tr	tr
15	16.35 ± 0.00	26.10 ± 1.11	33.05 ± 0.85	38.02 ± 1.35	24.10 ± 0.17	52.15 ± 0.75	19.55 ± 3.22	tr
16	1.48 ± 0.36	17.15 ± 0.00	21.21 ± 0.40	23.90 ± 0.35	19.11 ± 0.91	15.35 ± 0.25	14.34 ± 1.05	15.45 ± 0.65
17	45.53 ± 3.85	21.15 ± 0.35	68.63 ± 0.92	76.91 ± 1.11	50.20 ± 2.70	51.45 ± 0.65	5.23 ± 0.20	14.95 ± 0.05
18	13.05 ± 0.15	92.90 ± 3.21	171.80 ± 9.55	183.35 ± 2.55	111.90 ± 0.75	5.82 ± 0.30	0.35 ± 0.01	2.25 ± 0.01
19	76.30 ± 2.40	21.21 ± 0.30	24.85 ± 0.25	37.42 ± 0.53	15.12 ± 0.15	tr	tr	tr
20	27.20 ± 1.35	112.35 ± 3.35	179.92 ± 2.85	275.60 ± 3.85	7.65 ± 0.12	12.03 ± 3.50	11.02 ± 1.82	12.85 ± 0.25
21	13.45 ± 0.05	40.91 ± 1.35	104.65 ± 3.75	162.55 ± 2.30	82.74 ± 1.50	68.95 ± 0.55	7.15 ± 1.15	29.8 ± 0.60
22	13.12 ± 0.05	22.25 ± 0.30	24.10 ± 0.15	36.65 ± 0.50	tr	tr	tr	tr
23	127.85 ± 5.75	19.13 ± 0.50	25.95 ± 0.70	40.02 ± 0.55	22.95 ± 0.32	16.11 ± 0.30	14.92 ± 1.30	15.35 ± 0.05
24	106.5 ± 4.80	134.03 ± 1.25	329.73 ± 5.55	493.55 ± 6.91	190.22 ± 0.35	29.35 ± 0.61	20.41 ± 2.05	11.15 ± 0.35
Total	**492.25 ± 3.80**	**918.00 ± 1.46**	**1525.75 ± 2.31**	**1956.85 ± 5.48**	**682.00 ± 2.53**	**282.60 ± 0.17**	**120.80 ± 3.48**	**111.15 ± 2.05**

mean values ± SD (n = 3). tr – traces under LOD (LOD - limit of detection).

Table 4. The content of flavonols in red wines enriched with Primula veris L. (mg/L).

Red Wine	Regional				Commercial			
Primula veris L. (%)	0	2.5	5	10	0	2.5	5	10
Flavonols								
1	tr	8.15 ± 0.00	14.80 ± 0.00	28.60 ± 0.00	4.95 ± 0.22	1.95 ± 0.00	15.65 ± 0.05	22.65 ± 0.45
2	tr	tr	tr	41.82 ± 0.55	tr	3.30 ± 0.11	24.05 ± 0.9	23.30 ± 0.00
3	tr	tr	tr	14.01 ± 0.15	tr	1.55 ± 0.11	10.95 ± 0.6	8.35 ± 0.05
4	tr	5.52 ± 0.25	10.92 ± 0.75	16.70 ± 0.55	18.70 ± 0.75	0.91 ± 0.05	5.45 ± 0.15	10.75 ± 0.15
5	tr	19.21 ± 0.65	39.30 ± 0.35	59.15 ± 0.95	9.62 ± 0.55	4.72 ± 0.00	29.65 ± 0.75	41.31 ± 1.71
6	1.3 ± 0.05	141.10 ± 8.25	203.51 ± 6.25	290.63 ± 4.85	7.35 ± 0.15	27.41 ± 1.50	138.71 ± 2.75	188.55 ± 4.62
7	tr	2.14 ± 0.31	5.62 ± 0.25	9.21 ± 0.22	7.11 ± 0.00	2.65 ± 0.30	6.82 ± 0.10	11.95 ± 0.45
8	1.5 ± 0.00	7.42 ± 0.05	12.45 ± 0.45	27.45 ± 0.10	16.51 ± 0.21	6.72 ± 0.50	10.50 ± 0.61	28.42 ± 0.75
9	0.95 ± 0.05	6.95 ± 0.05	13.00 ± 0.05	21.75 ± 1.41	16.13 ± 0.20	3.33 ± 0.05	11.45 ± 0.10	9.40 ± 0.02
10	0.45 ± 0.00	36.81 ± 0.05	65.02 ± 0.55	118.11 ± 0.15	3.45 ± 0.10	11.35 ± 0.25	48.55 ± 0.75	53.95 ± 0.15
11	0.95 ± 0.05	49.60 ± 0.23	87.05 ± 0.85	145.72 ± 2.62	5.25 ± 0.05	15.85 ± 0.05	64.85 ± 2.35	83.82 ± 0.30
12	4.5 ± 0.05	282.42 ± 1.25	520.05 ± 0.05	892.30 ± 0.50	tr	89.85 ± 3.25	359.21 ± 2.40	469.71 ± 8.20
13	tr	14.12 ± 0.2	14.35 ± 0.10	51.21 ± 0.35	tr	7.85 ± 0.11	21.61 ± 0.60	37.10 ± 0.00
14	2.35 ± 0.01	14.30 ± 0.3	21.75 ± 0.05	48.75 ± 0.15	tr	21.75 ± 0.95	45.25 ± 0.75	131.55 ± 0.40
15	16.25 ± 1.01	4.75 ± 0.15	6.75 ± 0.11	18.85 ± 0.00	tr	2.20 ± 0.10	6.72 ± 0.11	13.02 ± 0.05
16	tr	7.55 ± 0.05	8.85 ± 0.01	29.65 ± 0.25	4.42 ± 0.10	55.25 ± 0.40	78.65 ± 4.0	100.52 ± 1.82
17	5.6 ± 0.55	1.23 ± 0.05	3.52 ± 0.32	13.15 ± 0.20	8.05 ± 0.15	17.71 ± 1.15	30.45 ± 0.51	38.05 ± 0.50
18	29.25 ± 2.04	11.85 ± 0.05	13.75 ± 0.00	56.62 ± 0.05	3.21 ± 0.05	92.25 ± 2.45	162.45 ± 6.73	136.83 ± 1.65
Total	**63.55 ± 4.01**	**613.00 ± 2.25**	**1040.55 ± 5.30**	**1883.61 ± 4.70**	**104.55 ± 4.14**	**366.50 ± 4.68**	**1070.90 ± 4.21**	**1409.05 ± 3.09**

mean values ± SD (n = 3). tr – traces under LOD.

Table 5. The content of flavonols in white wines enriched with *Primula veris* L. (mg/L).

White Wine	Regional				Commercial			
Primula veris L. (%)	0	2.5	5	10	0	2.5	5	10
Flavonols								
1	tr	3.02 ± 0.00	5.65 ± 0.01	16.95 ± 0.12	tr	19.55 ± 0.50	24.21 ± 0.05	29.82 ± 0.50
2	tr	tr	tr	39.21 ± 0.50	0.42 ± 0.02	52.05 ± 4.05	65.41 ± 1.55	78.65 ± 1.50
3	tr	tr	tr	22.75 ± 0.65	tr	32.21 ± 0.45	30.60 ± 0.05	36.45 ± 0.25
4	tr	5.01 ± 0.31	5.25 ± 0.10	9.51 ± 1.02	tr	10.12 ± 0.11	12.10 ± 0.11	12.75 ± 0.45
5	tr	10.70 ± 0.25	19.12 ± 0.22	33.62 ± 0.11	0.45 ± 0.02	43.25 ± 0.45	45.21 ± 0.45	54.92 ± 1.02
6	39.50 ± 0.05	13.17 ± 0.85	85.60 ± 3.35	194.45 ± 10.75	3.92 ± 0.11	239.50 ± 4.21	365.15 ± 3.00	411.91 ± 1.25
7	4.35 ± 0.15	2.33 ± 0.01	4.72 ± 0.12	8.50 ± 0.21	4.71 ± 0.20	13.72 ± 0.15	8.55 ± 0.01	10.85 ± 0.01
8	16.22 ± 0.65	7.05 ± 0.25	28.71 ± 0.12	58.85 ± 2.75	17.10 ± 0.25	55.85 ± 0.05	75.02 ± 3.25	85.22 ± 0.15
9	4.45 ± 0.05	2.95 ± 0.00	9.83 ± 0.05	20.03 ± 0.22	4.61 ± 0.10	14.25 ± 0.05	15.85 ± 1.15	16.45 ± 0.00
10	1.55 ± 0.1	26.6 ± 0.00	36.80 ± 1.45	83.45 ± 1.70	2.12 ± 0.02	83.02 ± 0.10	136.31 ± 2.05	152.70 ± 0.95
11	1.60 ± 0.00	27.65 ± 0.05	105.62 ± 1.62	15.51 ± 0.15	2.21 ± 0.00	115.05 ± 6.01	126.32 ± 0.55	151.12 ± 2.52
12	1.92 ± 0.15	105.81 ± 0.40	527.01 ± 0.35	890.85 ± 2.35	11.95 ± 0.25	717.75 ± 4.8	1044.75 ± 5.15	1086.85 ± 2.05
13	10.65 ± 0.45	23.42 ± 0.32	48.05 ± 0.50	97.75 ± 1.05	1.15 ± 0.02	54.35 ± 2.21	69.20 ± 0.05	79.70 ± 1.35
14	0.45 ± 0.00	28.35 ± 0.05	81.65 ± 0.00	126.72 ± 1.05	3.62 ± 0.05	62.45 ± 5.45	75.25 ± 4.05	80.55 ± 1.81
15	1.95 ± 0.11	1.75 ± 0.00	5.55 ± 0.05	9.70 ± 0.05	1.70 ± 0.00	10.95 ± 0.42	20.04 ± 0.15	20.61 ± 0.21
16	4.21 ± 0.01	15.75 ± 0.15	165.51 ± 1.50	197.75 ± 3.75	4.55 ± 0.05	49.80 ± 0.65	26.82 ± 0.55	24.65 ± 0.12
17	1.25 ± 0.02	5.41 ± 0.71	23.15 ± 1.15	31.35 ± 7.50	1.45 ± 0.01	18.55 ± 0.90	9.20 ± 0.05	11.95 ± 0.05
18	4.75 ± 0.15	21.70 ± 0.10	285.02 ± 2.05	360.31 ± 1.12	4.52 ± 0.11	77.85 ± 0.45	46.95 ± 1.22	59.25 ± 1.13
Total	**92.50 ± 1.66**	**300.80 ± 0.52**	**1434.80 ± 2.82**	**2356.80 ± 2.18**	**64.35 ± 1.14**	**1668.90 ± 1.18**	**2197.25 ± 3.34**	**2404.30 ± 2.38**

mean values ± SD (n = 3), tr – traces under LOD.

The assessment of the pro-health properties of wines enhanced with *Primula veris* L. flowers is also based on quantitative and qualitative assay of saponins in red and white wines, the results of which are shown in Table 6.

Table 6. The content of saponins in red and white wines enriched with *Primula veris* L. (mg/L).

Primula veris L. (%)	Regional			Commercial		
	Red Wine					
	2.5	5	10	2.5	5	10
Saponins						
1	28.2 ± 0.29	43.35 ± 0.15	88.75 ± 0.23	25.30 ± 0.05	31.40 ± 0.25	64,70 ± 0.19
2	111.4 ± 0.42	161.85 ± 0.50	318.90 ± 1.93	75.65 ± 0.33	134.10 ± 0.69	222.90 ± 0.26
3	9.35 ± 0.07	16.55 ± 0.13	30.65 ± 0.29	2.90 ± 0.01	5.70 ± 0.05	10.00 ± 0.11
Total	148.95 ± 0.78	221.70 ± 0.78	438.35 ± 2.45	103.90 ± 0.27	171.25 ± 1.00	297.60 ± 0.56
	White Wine					
1	23.00 ± 0.09	34.70 ± 0.39	51.15 ± 0.31	24.80 ± 0.06	39.80 ± 0.34	52.20 ± 0.27
2	102.45 ± 0.05	124.00 ± 0.50	226.05 ± 1.71	87.15 ± 0.12	118.70 ± 0.96	199.35 ± 0.41
3	6.35 ± 0.07	8.70 ± 0.01	22.45 ± 0.06	8.25 ± 0.04	17.65 ± 0.02	19.25 ± 0.16
Total	131.75 ± 0.21	167.40 ± 0.90	299.65 ± 2.07	120.20 ± 0.10	173.20 ± 1.32	270.80 ± 0.01

mean values ± SD (n = 3).

It was observed in all the test samples of enhanced wines that the substance present in the greatest quantities was Primula saponin I, particularly in the regional red wines. On the other hand, the lowest content was observed for Priverosaponin B 22-acetate, in particular in the commercial red wines enhanced with *Primula veris* L. According to multiple authors, this substance is a saponin characteristic of *Primula veris* L. roots [13]. The total saponin content in the regional red wine blends was 145.70 mg/L at the 2.5% addition of *Primula veris* L. flowers, while at the 10% flower addition, the saponin content was 438.35 mg/L. In the red commercial wines enhanced with the lowest *Primula veris* L. addition, on the other hand, the total saponin content was found to be 103.90 mg/L, while at the highest flower addition, the total content of these substances increased almost three-fold. In the case of the commercial white wines, the 2.5% cowslip flower addition enhanced the wine blends with 120.20 mg/L of saponins, while at the 10% flower addition, the total content of these substances was 270.80 mg/L. The regional white wines were characterized by an ability to extract saponins to solutions on a similar level to the regional red wines, i.e., 131.75 mg/L, while when enhanced with 10% *Primula veris* L., the saponin content increased to 299.65 mg/L.

Considering the varied extraction ability of the white and red wines in relation to the pro-health substances from *Primula veris* L., the impact of ethanol strength on extraction when ethanol was used to extract polyphenolic substances and saponins from *Primula veris* L. flowers that were added to regional and commercial wines. An analysis of the results, which is shown in Table 7, confirmed the adverse effect of ethanol strength on the extraction of polyphenolic compounds from *Primula veris* L. flowers.

Table 7. The content of flavonols and saponins in ethanol solvents enriched with *Primula veris* L. (mg/L).

Primula veris L. (%)		5	
Etanol (%)	**40**	**70**	**96**
		Flavonols	
1	tr	tr	tr
2	tr	tr	tr
3	tr	tr	tr
4	tr	tr	tr
5	1.35 ± 0.02	tr	tr
6	7.30 ± 0.01	1.05 ± 0.02	0.20 ± 0.02
7	tr	tr	tr
8	2.50 ± 0.27	0.55 ± 0.01	0.30 ± 0.00
9	0.80 ± 0.00	tr	tr
10	3.80 ± 0.00	0.45 ± 0.00	0.20 ± 0.00
11	5.25 ± 0.00	0.80 ± 0.00	0.30 ± 0.00
12	24.4 ± 0.15	3.60 ± 0.01	1.80 ± 0.02
13	2.05 ± 0.00	0.25 ± 0.00	0.40 ± 0.01
14	11.20 ± 1.63	7.55 ± 0.04	0.40 ± 0.01
15	tr	tr	tr
16	27.95 ± 0.06	4.25 ± 0.03	2.70 ± 0.01
17	6.30 ± 0.01	1.20 ± 0.00	0.55 ± 0.00
18	52.75 ± 1.64	11.25 ± 0.05	3.90 ± 0.00
Total	145.70 ± 3.09 ^{C}	27.65 ± 0.01 ^{B}	11.00 ± 0.01 ^{A}
		Saponins	
1	34.35 ± 0.36 ^{A}	37.65 ± 0.08 ^{B}	93.15 ± 0.40 ^{C}
2	117.90 ± 0.44 ^{A}	167.65 ± 0.35 ^{B}	317.65 ± 0.77 ^{C}
3	16.25 ± 0.13 ^{A}	23.25 ± 0.12 ^{B}	33.00 ± 0.07 ^{C}
Total	168.50 ± 0.92 ^{A}	228.55 ± 0.31 ^{B}	443.80 ± 0.44 ^{C}

mean values ± SD (n = 3), tr – traces under LOD; Statistically significant differences between means ($^{A–C}$ for $p \leq 0.01$), marked by different letters in the rows.

In samples with the lowest ethanol strength (40%), the highest mean flavonol content of 145.70 mg/L among all ethanol samples was achieved, while for the highest ethanol strength (95%), the value was reduced to 11.00 mg/L. The results of analysis of saponin content in ethanol blends also confirmed the statistically highly significant differences between the mean saponin content in ethanol solutions, with the relation being reverse to that in the case of flavonols. Ethanol strength had a beneficial impact on saponin extraction from cowslip flowers. In samples with the lowest ethanol strength (40%), a concentration of 168.50 mg/L was found, while at the highest ethanol strength, more than 2.5 times more saponins were found than for the lowest value. As in the case of wines enhanced with *Primula veris* L. flowers, Primula saponin I was extracted in the greatest amounts (117.90–317. 65 mg/L), while Priverosaponin B22 was present in the smallest quantities (16.25–33.00 mg/L).

Table 8 shows the results of statistical analysis of the influence of type of wine and the level of *Primula veris* L. addition on the profile of polyphenolic compounds and antioxidant properties, ethanol, and sugars content in enhanced wines.

Table 8. Influence of type of wine and the level of *Primula veris* L. addition on the profile of polyphenolic compounds and antioxidant properties, ethanol, and sugars content in enriched wines. ABTS: 2,2-azino-bis(3-ethylbenzothiazoline-6-sulfonic acid), FRAP: ferric reducing antioxidant power method, TPC: total phenolic compounds.

		WINES				*Primula veris* L. (%)			
		RED		WHITE		0	2.5	5	10
		Regional	Commercial	Regional	Commercial				
mg/L	**Total Anthocyanins**	1223.20 ^{B}	308.20 ^{A}	—	—	290.49 ^{A}	300.14 ^{B}	423.14 ^{C}	517.00 ^{D}
	Total Flavonols	900.18 ^{B}	737.75 ^{A}	1037.40 ^{C}	1591.70 ^{D}	79.73 ^{A}	737.30 ^{B}	1435.90 ^{C}	2013.40 ^{D}
	Total Saponins	202.25 ^{C}	143.19 ^{A}	149.70 ^{B}	141.05 ^{A}	—	126.20 ^{A}	183.38 ^{B}	326.60 ^{C}
	FRAP	14.50 ^{D}	10.47 ^{C}	2.15 ^{A}	2.44 ^{B}	5.07 ^{A}	6.76 ^{B}	8.25 ^{C}	9.50 ^{D}
	DPPH	11.76 ^{D}	8.82 ^{C}	1.67 ^{A}	1.82 ^{B}	4.58 ^{A}	5.32 ^{B}	6.74 ^{C}	7.46 ^{D}
	ABTS	22.58 ^{D}	16.67 ^{C}	3.08 ^{A}	3.67 ^{B}	9.04 ^{A}	10.27 ^{B}	12.76 ^{C}	1393 ^{D}
	TPC	1577.80 ^{D}	932.35 ^{C}	431.51 ^{B}	359.64 ^{A}	482.97 ^{A}	726.62 ^{B}	898.38 ^{C}	1193.30 ^{D}
	Ethanol (%*v*/*v*)	10.14 ^{D}	9.91 ^{B}	10.56 ^{C}	9.38 ^{A}	11.33 ^{D}	10.34 ^{C}	9.31 ^{B}	8.97 ^{A}
g/L	**Glucose**	0.11 ^{A}	0.27 ^{C}	0.33 ^{D}	0.23 ^{B}	0.08 ^{A}	0.33 ^{D}	0.23 ^{B}	0.29 ^{C}
	Fructose	0.33 ^{B}	0.21 ^{A}	0.43 ^{C}	0.34 ^{B}	0.18 ^{A}	0.54 ^{D}	0.26 ^{B}	0.32 ^{C}
Total Sugars		0.43 ^{A,a}	0.48 ^{b}	0.75 ^{D}	0.56 ^{C}	0.26 ^{A}	0.87 ^{D}	0.48 ^{B}	0.61 ^{C}

Statistically significant differences between means ($^{A–D}$ for $p \leq 0.01$; $^{a–b}$ for $p \leq 0.05$), marked by different letter in the rows.

Statistical analysis demonstrated highly significant differences between anthocyanin content in the regional red wines, which were characterized by a four times higher content of these substances compared to the commercial wines. Furthermore, statistically highly significant differences between the total saponin content in the enhanced wine samples were confirmed. The red wines displayed a 25% higher saponin content as compared to the remaining wine samples in this respect. Additionally, the regional red wines also showed the highest antioxidant potential, as determined using the ferric reducing antioxidant power method (FRAP), DPPH, and 2,2-azino-bis(3-ethylbenzothiazoline-6-sulfonic acid) (ABTS), as well as total phenolic compounds (TPC) methods. Lower antioxidant ability assay values were found for white wines, particularly the regional ones, which were, on the other hand, characterized by the highest sugar content among all the test samples. The commercial white wines displayed the highest flavonol content. Cowslip enhancement resulted in statistically highly significant differences in each of the analyzed ingredients of white and red wines. The highest flower content of 10% markedly improved the content of anthocyanins in wines by approximately 80%, the flavonol content as compared to wines without cowslip addition by up to 28 times, and the saponin content almost 2.5 times as compared to the lowest *Primula veris* L. flower addition. Cowslip addition also resulted in a statistically high improvement of the anti-oxidant ability of the test wines, as determined using the FRAP and DPPH, and particularly ABTS methods, increasing this value more than 130 times. Wine enhancement with cowslip resulted in a reduction in alcohol content, and a slight increase in sugar content. However, this did not significantly affect wine flavor or its classification, and the wines remained in the dry wine segment. The content of residual sugar in dry wines ranges between 2 and 4 g/L [4], meaning that all the tested wines were classified as dry.

3. Discussion

Over 1000 mineral and organic compounds have been discovered in wines to date. The content of these compounds depends on the following factors: grapevine variety, climate and soil conditions, and grape ripeness [3].

Assessing the quality of wines enhanced with herbal plants that were analyzed in this paper is difficult due to the lack of literature data on this subject. There is also little data in the scientific literature concerning the quality of Polish wines. Analyzing reports on the relation between the pro-health quality of wines and their method of production, one can find information, for example, that red wines contain 10 times as much polyphenolic compounds as white wines [18]. This mainly results from their production method. The polyphenol content in wine depends mainly on the maceration process and its duration. If the grapes are pressed too quickly, a light-colored grape must forms, as a large group of polyphenolic compounds, e.g., anthocyanins, is located in grape skin. Polyphenolic compounds have a major influence on such properties as color, sharpness, and bitterness, and are therefore important indicators in grapevine cultivation and enology. A particularly important role in the quality of red grapes and wine is played by color anthocyanins, whose peculiar profile is the main indicator for classifying grape and wine varieties [19].

In general, anthocyanin content in wines produced from interspecies hybrid grapes (e.g., the tested regional wine Rondo) is higher and ranges from 400 to 700 mg/L, and in some cases may exceed 2000 mg/L [20,21]. In traditional wine production, anthocyanin concentration may change after a few days of fermentation, due to the absorption of certain anthocyanins by yeast cell walls, and precipitation as wine salts; therefore, the anthocyanin profile even in wines made of the same grapevine varieties may significantly differ [22]. One should mind that the grapevines cultivated in Poland are a relatively new material of poor quality [20–25]. Samoticha et al. [26] demonstrated that the highest concentration of anthocyanins occurred in the Regent and Rondo wines produced in Poland, but the content of their derivatives depends on the grape variety as well as the winemaking technique.

In this study, flavonol profiles concerning white wine characteristics contained significantly more derivatives than those reported in the studies that are available in the literature, as other authors very frequently mention only the major compounds, such as myricetin glycosides, quertecin, and to a lesser extent kaempferol [27]. According to the data shown by Makris et al. [28], the flavonol content in white wines produced from various grape varieties and in different regions may range from 2 to 7 mg/L. In this study, flavonol content was on average twice as high as in wines from other regions. In the case of red wines, flavonol content can range from 5 mg/L to 100 mg/L, according to literature data. It is believed that the flavonoids present in red wine are responsible for so called "French paradox", which involves a low occurrence of cardiovascular diseases in populations with high red wine intakes [29].

The determination of ethyl alcohol content is of high importance in the winemaking industry. This parameter is used to control the fermentation process and for certification of alcoholic and non-alcoholic products. It is the main organic by-product of fermentation performed by yeast [30]. Most countries require stating ethanol content on labels, and moreover, wine price depends on the content of this compound in some cases [3]. The low ethanol content in Polish wines—9.7% to 11.9% *v*/*v* on average, and 12.5% *v*/*v* for the Rondo variety—has been observed by Tarko et al. [9,31]. On the other hand, in the Kapusta et al. [4] study, ethanol content was 12.44% *v*/*v* on average in red wines, and 12.30% *v*/*v* in white wines. The results concerning ethanol content in the test red and white wine samples in this study were similar to the literature data and consistent with the manufacturers' information on the labels. The alcohol content was reduced in samples enhanced with *Primula veris* L.

Sugar content in grapes depends mainly on the variety and on fruit ripeness. It is also affected by fermentation microflora, as mentioned previously. In the Polish climate, the ripening process runs in a manner largely different than in warmer regions. It results in a significantly lower sugar content, which may also translate to a lower ethanol content [32]. The main sugars present in grapes are the monosaccharides glucose and fructose, which are key for the growth and development of yeast. When the fermentation process is complete, so-called residue sugar remains, which is made up of pentose sugars (arabinose, rhamnose, xylose) and a small quantity of unfermented fructose and glucose [3].

The antioxidant properties of wines vary greatly depending on the grapevine species as well as environmental and geographical factors. They result in particular from the content of biologically active compounds, such as phenolic compounds, vitamins, and enzymes. As has been noted previously, red wines are a richer source of polyphenols and display relatively higher antioxidant properties in comparison to white wines. In studies performed by other researchers, it has been demonstrated that the varieties analyzed in this study, i.e., Rondo, are particularly rich in polyphenolic compounds and display a high antioxidant activity [5]. In the study of Kapusta et al. [4], who analyzed wine samples from Subcarpathian vineyards, antioxidant activity measured by the FRAP method ranged from 2 mmolTE/L (Cabernet Cortis variety) to 0.05 mmolTE/L (Regent variety); the Regent variety also contained the most polyphenolic compounds as assayed using LC/MS (1860 mg/L). The Rondo wines were also characterized by a very high content of polyphenolic compounds in the Socha et al. [33] study, 996 and 1669 mg GA/L, respectively. Furthermore, Socha et al. [33] observed that wines from the southern regions of Poland have a similar content of polyphenolic compounds to wines from regions with continental climates, but lower than red wines from warm regions, such as Greece, Portugal, or Italy.

A study conducted in recent years in Southeastern European countries demonstrated a rich biovariety and exceptional vitality of traditional plant knowledge in that region [34,35]. Familiarity with cowslip (*Primula veris* L.) in that part of Europe is not limited merely to applications in traditional medicine; it is also used as an additive in food [36]. This traditional plant product is widely used in several Central European countries in herbal preparations and pharmaceutical formulations, and its biological and pharmacological activity has been confirmed both in scientific and medical literature. Cowslip (*Primula veris* L., syn. *P. officinalis* Hill) and oxlip (*Primula elatior* (L.) Hill) are small, long-lived perennials from the family Primulaceae, growing wild in Europe and Asia [37]. Both species have a long history of medicinal use. In the current issue of the European Pharmacopoeia, they are listed as a source of *Primula* root, from which bioactive substances for pharmacological applications are acquired [38]. However, in the British Herbal Pharmacopoeia [39], as well as the Pharmacopee Francaise [40], only *Primula veris* L. is listed as a source of raw material for the production of pharmaceutical and pro-health preparations with mucoactive, anti-inflammatory, diuretic, antimicrobial, antifungal, and sedative effects [16,41,42].

The main ingredient used by the pharmaceutical industry from *Primula veris* L. is saponins, whose content in the plant's roots may reach 5–10%. Saponins from *Primula veris* L. are triterpenoid glycosides with the oleanane ring system linked to carbohydrate moieties [43].

However, the authors of other studies note difficulties in the precise qualitative determination of the content of saponins extracted from *Primula veris* L. roots, which is most likely a consequence of a lack of chromophore groups, impairing UV detection. Due to the great genetic variation and the significance of climate, soil, and geography conditions in determining the content of bioactive substances in *Primula veris* L., a clear determination of the profile of pro-health properties for cowslip is extremely difficult [44,45]. Many authors focus their attention primarily on the roots of *Primula veris* L., considering the potential concentration of bioactive substances in these morphological parts [46]. Other researchers, basing on the latest analytical techniques, question the profiles of bioactive compounds from Primula veris L., as determined previously [14].

According to own research, *Primula veris* L. flowers are equally valuable as an object of research, as they are rich in bioactive substances, and their extraction does not require complex preparative techniques.

In the latest study performed by Apel et al. [47], who used high performance liquid chromatography/diode array detector/mass spectrometry (HPLC-DAD-MS) to study the methanol extracts from the petals of three Primula varieties, including *Primula veris* L., a broad spectrum of flavonoids and their glycosylated or methylated derivatives has been shown, which is consistent with the results obtained in our study. Furthermore, these researchers demonstrated that extracts from the leaves of three *Primula* samples were characterized by highly similar flavonoid and saponin profiles to the leaves; however, individual compounds were present in different relative quantities. The root extracts of all three Primula samples provided identical compound profiles, which consisted of triterpenoid saponins, but were devoid of the flavonoids and anthocyanins detected in the above-ground plant organs. Among saponins, the highest concentration in the test extracts was observed for Primula saponin I, which our study has confirmed as well. According to the study by Müller et al. [17], who tested methanol extracts obtained from dried roots and flowers of *Primula veris* L., the saponin dominant in the flowers is Primula saponin I, whose content can range from 0.22% to 0.28%. On the other hand, priverosaponin B-22-acetate has been identified only in root extracts, which according to the authors confirms the earlier reports concerning the saponin profile in various parts of the plant and contradicts the results obtained in our study [17]. Considering the above reports, it can be concluded that the content of biologically active substances in *Primula veris* L. has not been unambiguously determined and still leaves hope for new applications for the pro-health substances contained in *Primula veris* L. In the Committee on Herbal Medicinal Products (HMPC) report [41], information can be found about unspecified saponin of the Primula root, which administered parenterally in a dose of 40 mg/kg inhibited the growth of Walker carcinoma in rats.

Herbal medicinal products appear to be very promising as they have a noticeable therapeutic effect and tend to be more harmless in comparison to the most synthesized medications.

The study by Latypova et al. [48] focused particular attention on the identification of the raw material composition of the *Primula veris* L. The object of the analysis was a solid herbal extract and its effects on the myocardial contractile function in animals with experimental CHF (chronic heart failure). The authors of the study found that a solid herbal extract obtained from *Primula veris* L. contained flavonoid aglycons, flavonoid glycosides, and polymethoxylated flavonoids. It was further shown that the studied herbal agent at a dose of 30 mg/kg had a cardioprotective effect, as evidenced by a smaller number of animals deaths, the lower level of CHF plasma markers, a higher increase in myocardial contraction, and relaxation rates as compared to the control group [48].

Junqing et al. [49] found that herbs with a high percentage of phenolic constituents stimulate the synthesis of vascular endothelial growth factor and blood vessel density, and exert cardioprotection through promoting angiogenesis in the animal models of myocardial infarction [49,50].

The above effects of the studied herbal medicinal product from *Primula veris* L. are likely to be of key importance for undertaking further studies in this area, with the aim of providing cardioprotective benefits. The cardioprotective effects of wine has so far been attributed to resveratrol, which is a substance that is contained mainly in red wines, and in minimal quantities in white wines. Thanks to wine enrichment in triterpenoid saponins and in polyphenolic compounds from *Primula veris* L. flowers, which are effectively extracted to wine under mild conditions, both white and red wines can constitute a highly pro-health component of diets, which is valuable in preventing chronic heart failure.

4. Materials and Methods

4.1. Plant Material and Experimental Conditions

The test material (named Regional) were Polish, regional Rondo dry red wines (n = 3) from the "Wierzchowina" vineyard (marked W on Figure 1), located in Łęki Dolne, near Pilzno, at the base of the Carpathians in Poland, and the Rezeda Polish white wines (n = 3) produced at the regional vineyard "Jasiel" in Jaslo (marked J on Figure 1) in the Subcarpathian region, Poland. Rondo is an interspecies hybrid from Germany that is characterized by good resistance to frost (below −20°C) as well as pests, popular, and recommended for red wine production. Rezeda, on the other hand, is a mixture of the Muskaris, Siegerrebe, and Solaris varieties. The wine samples were stored in the dark, at a temperature of 4 °C, and opened directly before preparing the herbal blends.

For the purpose of comparing selected properties of the Subcarpathian regional products, red *Cabernet Sauvignon* (n = 3) and white *Chardonnay* (n = 3) wines, commercially available under the Carlo Rossi brand, were selected as control samples (named Commercial). Furthermore, fresh *Primula veris L* inflorescences were used in the tests. The test material were inflorescences of the cowslip (*Primula veris* L.) plant, which were collected at the beginning of April 2018 from environmentally friendly herb cultures in the Subcarpathian region, in Jasielski powiat. Washed and dried inflorescences were divided into portions that formed the charge for the wine blends in the amounts of 2.5%, 5%, and 10%, which were then immersed in white and red wine (regional and commercial) in broad-necked glass bottles. Subsequently, the samples were capped and left at room temperature (25 °C) in an air-conditioned room for a period of 14 days. The wine blends were stirred every day by vigorously turning the closed bottles upside down 3 times. Additionally, alcohol solvents of different alcohol strengths were prepared (40%, 70%, and 96%) using ethanol and distilled water, together with equal additions of *Primula veris* L. in the amount of 5% for this purpose. The samples were prepared in closed glass bottles and handled identically as the wine blends. Once the 14-day free extraction was complete, the wine samples were filtered and prepared directly for further analyses. Before analyzing the bioactive profiles, the wine and ethanol samples were filtered through paper filters at vacuum. Clear wine and alcohol solutions were applied to conditioned column beds of a Thermo Scientific™ SPE 16- or 24-Port SPE Vacuum Manifolds filtration system (C_{18} 500mg bed). Polyphenolic compounds were eluted from the columns using ethanol directly to round-bottom flasks, and concentrated at a temperature of 40 °C in a Hei-VAP Precision rotary vacuum evaporator manufactured by Heidolph (Germany), until complete evaporation of the solvent. The wine and alcohol extracts in round-bottom flasks were dissolved in methanol and filtered using PTFE filters with 0.45-μm pore diameter immediately before the chromatographic analysis.

4.2. Determination of Polyphenolic Compounds

The analysis was performed according to the method described by Kapusta et al. [51]. Polyphenolic compounds were analyzed using ultra-performance reverse-phase liquid chromatography (UPLC)-PDA-MS/MS Waters ACQUITY system (Waters, Milford, MA, USA), consisting of a binary pump manager, sample manager, column manager, photodiode array (PDA) detector, and tandem quadrupole mass spectrometer (TQD) with electrospray ionization (ESI). The separation was carried out using a BEH C18 column (100 mm × 2.1 mm i.d., 1.7 µm, Waters) kept at 50 °C. For the anthocyanins investigation, the following solvent system: mobile phase A (2% formic acid in water *v*/*v*) and mobile phase B (2% formic acid in 40% ACN in water *v*/*v*) were applied. For other polyphenolic compounds, a lower concentration of formic acid was used (0.1% *v*/*v*). The gradient program was set as follows: 0 min 5% B, from 0 to 8 min linear to 100% B, and from 8 to 9.5 min for washing and back to initial conditions. The injection volume of the samples was 5 µL (partial loop with needle overfill), and the flow rate was 0.35 mL/min. The following parameters were used for TQD: capillary voltage, 3.5 kV; con voltage, 30 V in positive and negative mode; the source was kept at 120 °C and the desolvation temperature was 350 °C; con gas flow, 100 L/h; and desolvation gas flow, 800 L/h. Argon was used as the collision gas at a flow rate of 0.3 mL/min. The polyphenolic detection and identification were based on a specific PDA spectra, mass-to-charge ratio and fragment ions obtained after collision-induced dissociation (CID). Before injection, wine samples were filtered through a 0.45-µm pore-size membrane filter (Merck Millipore) and injected directly into a chromatographic column. Quantification was achieved by the injection of solutions of known concentrations ranging from 0.05 to 5 mg/mL ($R^2 \leq 0.9998$) of phenolic compounds as standards. All determinations were performed in triplicate and expressed as mg/L. Waters MassLynx software v.4.1 was used for data acquisition and processing.

4.3. Determination of Saponins

Saponins were quantitated using ultra-performance reverse-phase liquid chromatography (UPLC) coupled with ESI-MS/MS described by Mroczek et al. [52]. Analyses were carried out using an Acquity ultra-performance liquid chromatograph (Waters) coupled with an Acquity TQD tandem quadrupole mass spectrometer with an ESI source. The separation was undertaken using a 100 × 2.1 mm i.d., 1.7 µm, Acquity UPLC BEH C_{18} column. A mobile phase consisting of 0.1% formic acid in acetonitrile (B) and 0.1% formic acid in water (A) was used for the separation. The gradient elution was linear from 25% to 60% B, over 0–6 min; isocratic at 60% B, over 6–6.5 min; linear from 60% to 25% B, over 6.5–6.6 min; and isocratic at 25% B, over 6.6–7 min. The column was maintained at 50 °C at a constant flow rate of 0.3 mL/min. The sample injection volume was 5 µL. The following instrumental parameters were used for the ESI-MS analysis of saponins: capillary voltage, 3.5 kV; cone voltage, 80 V; desolvatation gas, nitrogen 800 L/h; cone gas, nitrogen 100 L/h; source temperature, 120 °C; desolatation temperature, 350 °C; and dwell time, 0.05 s. The detection mode was SIR in negative ion mode. Quantitation was based on external standardization by employing the calibration curve of oleanolic acid in the range of 80–560 ng/mL. The quantitative analyses were based on the peak area calculated from the selected ion chromatograms of the corresponding $[M - H]^-$ ion, and saponins were identified through a comparison of their retention times and ion mass. Microsoft Excel 2000 was used for statistical analysis.

4.4. Analysis of Sugar Profiles

The sugar characteristics of wines were assessed by HPLC according to the method described by Dżugan et al. [53] with our own modifications. The sugar profile of samples was analyzed by high-performance liquid chromatography using a Thermo Dionex Ultimate 3000 with a charged aerosol detector (CAD) and ultraviolet-diode array detector (UV/DAD) (Thermo Fisher Scientific Inc., Waltham, MA, USA). Chromatographic separation was performed using a Shodex Asahipak NH2P-504E column (C18, 250 nm × 4.6 nm) and acetonitrile:water (78:22, *v*/*v*) as the mobile phase (1 mL/min). The injection volume was 10 µL, the temperature was set at 35 °C, and the analysis time was 30 min. Before injection, the samples were filtered using PTFE filters (0.22 µm). Sugars were identified on the basis of their retention times. The average yield for sugars in wines was 90–94%. The precision of this analytical method was confirmed by repeating the standard and all sample injections three times.

Chromatographic system stability was controlled in four-hour intervals by injecting selected standard solutions of known concentration.

4.5. Analysis of Antioxidant Activity

Antioxidant activity was measured by two different method: The ferric reducing antioxidant power method (FRAP) and radical scavenging activity (DPPH). Ferric reducing antioxidant power (FRAP) was performed according to Benzie and Strain [54]. The results were expressed as mmol of Trolox equivalents per 1 L of tested samples (mmol TE /L), using a calibration curve plotted for Trolox solution in a concentration range of 0.05–0.30 μmol/mL (R^2 = 0.9967). Scavenging activity was determined according to the elimination of DPPH radicals [55].

Antiradical activity was carried out using the ABTS decolorization assay described previously by Re et al. [56]. Determinations were performed in triplicate, and results were expressed as mmol of ascorbic acid equivalent per L. The determination of ABTS, DPPH, and FRAP methods was performed using a Spectrophotometer UV VIS UV6000 (Shanghai Metash Instruments Co., Shanghai, China).

4.6. Analysis of Total Phenolic Compounds (TPC)

The content of total phenolic compounds (TPC) was investigated by Folin–Cocialteu's method as described by Stratil et al. [57]. The results were expressed as 1 mg of gallic acid equivalents per l (mg GAE/L) of tested samples, using a calibration curve plotted for GAE solution in a concentration range of 25–250 μg/mL (R^2 = 0.997). The measurements were performed using a Spectrophotometer UV VIS UV6000 (Shanghai Metash Instruments Co., Shanghai, China).

4.7. Determination of Ethanol Content

The ethanol content in the wine samples was determined using the colorimetric method according to the methodology developed by Gonchar et al. [58], using the Alkotest analytical system, containing alcohol oxidase, peroxidase, chromogen, and buffering components with some modifications. The reaction was interrupted when the samples turned light blue by adding 100 μL of 0.8 M hydrochloric acid. Absorbance was measured at the 450 nm wavelength against a blank sample. The alcohol concentration results were read using a calibration chart, and then calculated in a manner similar to that shown in the Gonchar et al. [58] paper, and expressed in % *v*/*v*.

4.8. Chemicals and Reagents

The determinations were performed using analytical grade reagents intended for liquid chromatography: hydrochloric acid, formic acid, ethanol by Sigma-Aldrich; methanol by J.T. Baker Malinckrodt Baker B.V. Holland. Acetonitryl CHROMASOLV® gradient grade, ≥99.9% by Sigma-Aldrich. Sugar analytical standards by BioXtra, ≥99% HPLC grade, were obtained from Sigma-Aldrich. Deionized water was obtained using a HLP 5P deionizer manufactured by Hydrolab Polska. Oleanolic acid standard, 2,2-azino-bis(3-ethylbenzothiazoline-6-sulfonic acid) (ABTS) diammonium salt, 6-hydroxy-2,5,7,8-tetramethylchroman-2-carboxylic acid (Trolox), 2,4,6-tri(2-pyridyl)-s-triazine (TPTZ), and 2,2-diphenyl-1-picrylhydrazyl (DPPH) were purchased from Sigma-Aldrich (Steinheim, Germany).

4.9. Statistical Analysis

All the analyses were made in three independent replications for each sample. The results were presented as an arithmetic mean ± standard deviation (SD). The findings were subjected to statistical analyses with the use of Statistica 13.1 software (StatSoft, Inc., Tulsa, OK, USA). The significant differences between the mean values were obtained by a two-way analysis of variance (ANOVA) followed by Duncan's multiple range test.

5. Conclusions

As our study indicates, ethanol content has a significant impact on the extraction of health-beneficial polyphenolic compounds, reducing its effectiveness, while it advantageously affects saponin extraction from *Primula veris* L. flowers, increasing their concentration at higher ethanol strength levels. From the perspective of pro-health and the cardioprotective use of cowslip extracts, it is much more beneficial to prepare them based on white or red wines, which are products of low alcohol content, as it does not impair extraction, and additionally enriches the blends with the particular profile of the pro-health compounds contained therein.

Considering the above reports and the wealth of polyphenolic compounds contained in the analyzed red and white wines enhanced with *Primula veris* L. flowers, further action should be taken to clearly determine the ranges of fortification with cowslip extracts, as well as develop pro-health preventive practices that make use of wines enhanced with *Primula veris* L. flowers.

Author Contributions: Important contributions to the design and preparation of the manuscript: M.T. Contributions to sample and analysis experiments: M.T., A.G., and I.K. Analysis of the experimental data: M.T., I.K., and A.G. Critical revision for important intellectual content: M.T., I.K., and C.P. All authors helped prepare the paper and approved the final version.

Funding: Project funded under the Minister of Science and Higher Education programme "Regionalna Inicjatywa Doskonałości" for the years 2019–2022, project no. 026/RID/2018/19, funding amount PLN 9 542 500.00.

Conflicts of Interest: The authors declare no conflict of interest.

References

1. Reynolds, A.G. Viticultural and vineyard management practices and their effects on grape and wine quality. In *Managing Wine Quality*; Woodhead Publishing Limited: Cambridge, UK, 2010; pp. 365–444. [CrossRef]
2. Markoski, M.M.; Garavaglia, J.; Oliveira, A.; Olivaes, J.; Marcadenti, A. Molecular Properties of Red Wine Compounds and Cardiometabolic Benefits. *Nutr. Metab. Insights* **2016**, *9*, 51–57. [CrossRef] [PubMed]
3. Ribéreau-Gayon, P.; Glories, Y.; Maujean, A.; Dubourdieu, D. Handbook of Enology Volume 2. In *The Chemistry of Wine, Stabilization and Treatments*, 2nd ed.; John Wiley & Sons, Ltd.: Hoboken, NJ, USA, 2006.
4. Kapusta, I.; Cebulak, T.; Oszmiański, J. Characterization of polish wines produced from the interspecific hybrid grapes grown in south-east Poland. *Eur. Food Res. Technol.* **2018**, *244*, 441–455. [CrossRef]
5. Kunicka-Styczyńska, A.; Czyżowska, A.; Rajkowska, K.; Dziugan, P. The Trends and Prospects of Winemaking in Poland. *Grape and Wine Biotechnol. Intech. Open* **2016**, 401–413. [CrossRef]
6. Council Regulation (EC) No 479/2008 of 29 April 2008 on the common organisation of the market in wine, amending Regulations (EC) No 1493/1999, (EC) No 1782/2003, (EC) No 1290/2005, (EC) No 3/2008 and repealing Regulations (EEC) No 2392/86 and (EC) No 1493/1999. *Off. J. Eur. Union L141*, 1–61.
7. Wawro, E. *Winnice w Polsce*; Multico: Warszawa, Poland, 2015.
8. Lachowicz, S. Enotourism on Podkarpacie. *Sci. Rev. Phys. Cult.* **2014**, *4*, 5–13.
9. Tarko, T.; Duda-Chodak, A.; Satora, P.; Sroka, P.; Gojniczek, I. Chemical composition of cool-climate grapes and enological parameters of cool-climate wines. *Fruits* **2014**, *69*, 75–86. [CrossRef]
10. Kumar, S.; Pandey, A.K. Chemistry and biological activities of flavonoids: An overview. *Sci. World J.* **2013**, 162750. [CrossRef] [PubMed]
11. Ahmed, S.I.; Hayat, M.Q.; Tahir, M.; Mansoor, Q.; Ismail, M.; Keck, K.; Bates, R.B. Pharmacologically active flavonoids from the anticancer, antioxidant and antimicrobial extracts of Cassia angustifolia Vahl. BMC Complement. *Altern. Med.* **2016**, *16*, 4603.
12. Działo, M.; Mierziak, J.; Korzun, U.; Preisner, M.; Szopa, J.; Kulma, A. The potential of plant phenolics in prevention and therapy of skin disorders. *Int. J. Mol. Sci.* **2016**, *17*, 160. [CrossRef] [PubMed]
13. Grosso, G.; Bella, F.; Godos, J.; Sciacca, S.; Del Rio, D.; Ray, S.; Galvano, F.; Giovannucci, E.L. Possible role of diet in cancer: Systematic review and multiple meta-analyses of dietary patterns, lifestyle factors, and cancer risk. *Nutr. Rev.* **2017**, *75*, 405–419. [CrossRef] [PubMed]

14. Colombo, P.S.; Flamini, G.; Rodondi, G.; Giuliani, C.; Santagostini, L.; Fico, G. Phytochemistry of European Primula species. *Phytochemistry* **2017**, *143*, 132–144. [CrossRef] [PubMed]
15. Okrslar, V.; Plaper, I.; Kovac, M.; Erjavec, A.; Obermajer, T.; Rebec, A.; Ravnikar, M.; Zel, J. Saponins in tissue culture of *Primula veris* L. *In Vitro Cell Dev. Biol. Plant.* **2007**, *43*, 644–651. [CrossRef]
16. Budzianowski, J.; Morozowska, M.; Wesołowska, M. Lipophilic flavones of Primula veris L. from field cultivation and in vitro cultures. *Phytochemistry* **2005**, *66*, 1033–1039. [CrossRef] [PubMed]
17. Müller, A.; Ganzera, M.; Stuppner, H. Analysis of phenolic glycosides and saponins in Primula elatior and Primula veris (primula root) by liquid chromatography, evaporative light scattering detection and mass spectrometry. *J. Chromatogr. A* **2006**, *1112*, 218–223. [CrossRef] [PubMed]
18. Mas, A.; Padilla, B.; Esteve-Zarzoso, B.; Beltran, G.; Reguant, C.; Bordons, A. Taking advantage of natural biodiversity for wine making: The WILDWINE Project. *Agric. Agric. Sci. Procedia* **2016**, *8*, 4–9. [CrossRef]
19. Kennedy, J.A. Grape and wine phenolics: Observations and recent findings. *Cienc. Investig. Agrar.* **2008**, *35*, 107–120. [CrossRef]
20. Kelebek, H.; Canbas, A.; Jourdes, A.; Teissedre, P.L. Characterization of colored and colorless phenolic compounds in Okuzgozu wines from Denizil and Elazig regions using HPLC-DAD-MS. *Ind Crop. Prod.* **2011**, *31*, 499–508. [CrossRef]
21. Burns, J.; Mullen, W.; Landrault, N.; Teissedre, P.L.; Lean, M.E.J.; Crozier, A. Variations in the profile and content of anthocyanins in wines made from Cabernet Sauvignon and hybrid grapes. *J. Agric. Food Chem.* **2002**, *50*, 4096–4102. [PubMed]
22. He, F.; Liang, N.N.; Mu, L.; Pan, Q.H.; Wang, J.; Reeves, M.J.; Duan, C.Q. Anthocyanins and their variation in Red Wines, I. Monomeric anthocyanins and their color expression. *Molecules* **2012**, *17*, 1571–1601. [CrossRef] [PubMed]
23. Gąstoł, M.; Domagała-Świątkiewicz, I.; Krośniak, M. Antioxidant capacity, polyphenol and mineral content of grapes grown in southern Poland. *Acta Hortic.* **2012**, *931*, 345–348. [CrossRef]
24. Leja, M.; Kamińska, I.; Kulczak, K. Antioxidative properties in grapes of selected cultivars grown in Poland. *Ecol. Chem. Eng. S* **2011**, *18*, 59–65.
25. Tarko, T.; Duda-Chodan, A.; Sroka, P.; Satora, P.; Jurasz, E. Physicochemical and antioxidant properties of selected polish grape and fruit wines. *Acta Sci. Pol. Technol.* **2008**, *7*, 34–45.
26. Samoticha, J.; Wojdyło, A.; Golis, T. Phenolic composition, physicochemical properties and antioxidant activity of interspecific hybrids of grapes growing in Poland. *Food Chem.* **2017**, *215*, 263–273. [CrossRef] [PubMed]
27. Baderschneider, B.; Winterhalter, P. Isolation and characterization of novel benzoates, cinnamates, flavonoids, and lignans from Riesling wine and screening for antioxidant activity. *J. Agic. Food Chem.* **2001**, *9*, 2788–2798. [CrossRef] [PubMed]
28. Makris, D.P.; Kallithraka, S.; Kefalas, P. Flavonols in grapes, grape products and wines: Burden, profile and influential parameters. *J. Food Compost. Anal.* **2006**, *19*, 396–404. [CrossRef]
29. Foppa, M.; Fuchs, F.D.; Preissler, L.; Andrighetto, A.; Rosito, G.A.; Duncan, B.B. Red wine with the noon meal lowers post – meal blood pressure: A randomized trial in centrally obese, hypertensive patients. *J. Stud. Alcohol* **2002**, *63*, 247–251. [CrossRef] [PubMed]
30. Jackson, R.S. Chemical Constituents of Grapes and Wine. *Princ. Appl.* **2014**, 347–426. [CrossRef]
31. Tarko, T.; Duda-Chodak, A.; Sroka, P.; Satora, P.; Jurasz, E. Polish wines: Characteristics of cool-climate wines. *J. Food Comp. Anal.* **2010**, *23*, 463–468. [CrossRef]
32. Kemp, B.; Pedneault, K.; Pickering, G.; Usher, K.; Willwerth, J. Red Winemaking in Cool Climates. *Red Wine Technol.* **2019**, 341–356. [CrossRef]
33. Socha, R.; Gałkowska, D.; Robak, J.; Fortuna, T.; Buksa, K. Characterization of Polish Wines Produced from the Multispecies Hybrid and Vitis vinifera L. Grapes. *Int. J. Food Properties.* **2015**, *18*, 699–713. [CrossRef]
34. Kołodziejska-Degórska, I. Mental herbals - A contextsensitive way of looking at local ethnobotanical knowledge: Examples from Bukovina (Romania). *Trames* **2001**, *16*, 287–301. [CrossRef]

35. Nedelcheva, A.; Dogan, Y.; Obratov-Petkovic, D.; Padure, I.M. The traditional use of plants for handicrafts in southeastern Europe. *Hum. Ecol.* **2011**, *39*, 813–828. [CrossRef]
36. Pieroni, A.; Nedelcheva, A.; Hajdari, A.; Mustafa, B.; Scaltriti, B.; Cianfaglione, K.; Quave, C. Local knowledge on plants and domestic remedies in the mountain villages of Peshkopia (Eastern Albania). *J. Mt. Sci.* **2011**, *11*, 180–193. [CrossRef]
37. Wichtl, M. (Ed.) Herbal Drugs and Phytopharmaceuticals. In *A Handbook of Practice on a Scientific Basis*, 3rd ed.; CRC Press: Stuttgart, Germany, 2004.
38. European Pharmacopoeia. "Primula root (*Primulaeradix*)". In *European Directorate for the Quality of Medicines and Health Care (EDQM)*, 5th ed; Council of Europe: Strasbourg, France, 2006; pp. 2310–2311.
39. British Herbal Pharmacopoeia. *Primula veris*; British Herbal Medicine Association: London, UK, 1974.
40. *Pharmacope'e Franc¸aise*, 11th ed.; L'Adapharm: Lion, France, 2016.
41. "Assessment report on *Primulaveris* L. and/or *Primulaelatior* (L.) Hill, radix". In *EMA/HMPC/113577/2012EMA*; EMA (European Medicines Agency): Amsterdam, The Netherlands, 2013.
42. Gamze, A.; Ozmen, H.; Biyik, H.; Ozge, S. Antimitotic and antibacterial effects of the Primula veris L. flower extracts. *Caryologia* **2008**, *61*, 88–91.
43. Coran, S.A.; Mulas, S. Validated determination of primula saponins in primula root by a highperformance-thin-layer-chromatography densitometric approach. *J. Pharm. Biomed. Anal.* **2012**, *70*, 647–651. [CrossRef] [PubMed]
44. Hahn-Deinstrop, A.; Koch, M. Muller, HPTLC measured values of primula saponins in extracts of Primulae radix. A comparison after derivatization between handspraying, dipping and autospraying. *Chromatographia* **2000**, *51*, S302–S304. [CrossRef]
45. Kazimierczak, R.; Hallmann, E.; Ardasińska, B.; Łoś, B.; Rembiałkowska, E. Wpływ ekologicznego i konwencjonalnego systemu uprawy na zawartość związków fenolowych w roślinach zielarskich. *J. Res. Appl. Agric. Eng.* **2002**, *57*, 198–203.
46. Colombo, P.S.; Flamini, G.; Christodoulou, M.S.; Rodondi, G.; Vitalini, S.; Passarella, D.; Fico, G. Farinose alpine Primula species: Phytochemical and morphological investigations. *Phytochemistry* **2014**, *98*, 151–159. [CrossRef] [PubMed]
47. Apel, L.; Kammerer, D.R.; Stintzing, F.C.; Spring, O. Comparative Metabolite Profiling of Triterpenoid Saponins and Flavonoids in Flower Color Mutations of *Primula veris* L. *Int. J. Mol. Sci.* **2017**, *18*, 153. [CrossRef] [PubMed]
48. Latypova, G.M.; Bychenkova, M.A.; Katayev, V.A.; Perfilova, V.N.; Tyurenkov, I.; Mokrousov, I.S.; Prokofiev, I.I.; Salikhov, S.H.M.; Iksanova, G.R. Composition and cardioprotective effects of Primula veris L. solid herbal extract in experimental chronic heart failure. *Phytomedicine* **2018**, *54*. [CrossRef] [PubMed]
49. Junqing, G.; Tao, C.; Huigen, J.; Zongjun, L.; Deqiang, Z. Effect of calycosin on left ventricular ejection fraction and angiogenesis in rat models with myocardial infarction. *J. Tradit. Chin. Med.* **2015**, *35*, 160–167. [CrossRef]
50. Yu, L.J.; Zhang, K.J.; Zhu, J.Z.; Zheng, Q.; Bao, X.Y.; Thapa, S.; Wang, Y.; Chu, M.P. Salvianolic acid exerts cardioprotection through promoting angiogenesis in Animals models of acute myocardial infarction: Preclinical evidence. *Oxid. Med. Cell Longev.* **2017**, 8192383. [CrossRef] [PubMed]
51. Kapusta, I.; Cebulak, T.; Oszmiański, J. The anthocyanins profile of red grape cultivars growing in southeast Poland (Subcarpathia region). *Food Meas.* **2017**, *11*, 1863–1873. [CrossRef]
52. Mroczek, A.; Kapusta, I.; Stochmal, A.; Janiszowska, W. MS/MS and UPLC-MS profiling of triterpenoid saponins from leaves and roots of four red beet (*Beta vulgaris* L.) cultivars. *Phytochem. Lett.* **2019**, *30*, 333–337. [CrossRef]
53. Dżugan, M.; Sowa, P.; Kwaśniewska, M.; Wesołowska, M.; Czernicka, M. Physicochemical Parameters and Antioxidant Activity of Bee Honey Enriched with Herbs. *Plant Foods Hum. Nutr.* **2017**, *72*, 74–81. [CrossRef] [PubMed]
54. Benzie, I.F.F.; Strain, J.J. Ferric reducing ability of plasma (FRAP) as a measure of antioxidant power: The FRAP assay. *Anal. Biochem.* **1996**, *239*, 70–76. [CrossRef] [PubMed]
55. Shimada, K.; Fujikawa, K.; Yahara, K.; Nakamura, T. Antioxidative properties of xanthone on the auto oxidation of soybean in cylcodextrin emulsion. *J. Agric. Food Chem.* **1992**, *40*, 945–948. [CrossRef]

56. Re, R.; Pellegrini, N.; Proteggente, A.; Pannala, A.; Yang, M.; Rice-Evans, C. Antioxidant activity applying an improved ABTS radical cation decolorization assay. *Free Rad. Biol. Med.* **1999**, *26*, 1231–1237. [CrossRef]
57. Stratil, P.; Kubáň, V.; Fojtová, J. Comparison of the Phenolic Content and Total Antioxidant Activity in Wines as Determined by Spectrophotometric Methods. *Czech. J. Food Sci.* **2008**, *26*, 242–253. [CrossRef]
58. Gonchar, M.V.; Maidan, M.M.; Pavlishko, H.M.; Sibirny, A.A. A New Oxidase-Peroxidase Kit for Ethanol Assays in Alcoholic Beverages. *Food Technol. Biotechnol.* **2001**, *39*, 37–42.

Sample Availability: Samples of the compounds are not available from the authors.

molecules

MDPI

Article

Determining High-Intensity Sweeteners in White Spirits Using an Ultrahigh Performance Liquid Chromatograph with a Photo-Diode Array Detector and Charged Aerosol Detector

Kang Ma [1,2,*], Xiaojia Li [3], Yiwen Zhang [4] and Fei Liu [2,*]

[1] Division of Chemical Metrology and Analytical Science, National Institute for Metrology of China, Beijing 100013, China
[2] Beijing Key Laboratory of Water Resources and Environmental Engineering, China University of Geosciences (Beijing), Beijing 100083, China
[3] College of Chemistry and Bioengineering, University of Science and Technology Beijing, Beijing 100083, China; lixiaojia_xszx@aliyun.com
[4] Beijing Institute of Metrology, Beijing 100021, China; zhangyw@bjjl.cn
* Correspondence: makang@nim.ac.cn (K.M.); feiliu@cugb.edu.cn (F.L.); Tel.: +86-010-64524783 (K.M.); +86-010-82321027 (F.L.); Fax: +86-010-6452478 (K.M.); +86-010-82321081 (F.L.)

Academic Editors: Rosa Maria de Sá Perestrelo and José Sousa Câmara
Received: 12 November 2019; Accepted: 14 December 2019; Published: 20 December 2019

Abstract: In China, white spirit is not only an alcoholic drink but also a cultural symbol. A novel and accurate method for simultaneously determining nine sweeteners (most authorized for use in China) in white spirits by ultrahigh performance liquid chromatography (UHPLC) with a photo-diode array detector (PDA) and charged aerosol detector (CAD) was developed. The sweeteners were acesulfame, alitame, aspartame, dulcin, neotame, neohesperidine dihydrochalcone, saccharin, sodium cyclamate, and sucralose. The sweeteners were separated within 16 min using a BEH C18 column and linear gradient-elution program. The optimized method allowed low concentrations (micrograms per gram) of sweeteners to be simultaneously detected. The CAD gave good linearities (correlation coefficients > 0.9936) for all analytes at concentrations of 0.5 to 50.0 μg/g. The limits of detection were 0.16 to 0.77 μg/g. Acesulfame, dulcin, neohesperidine dihydrochalcone, and saccharin were determined using the PDA detector, which gave correlation coefficients > 0.9994 and limits of detection of 0.16 to 0.22 μg/g. The recoveries were 95.1% to 104.9% and the relative standard deviations were 1.6% to 3.8%. The UHPLC-PDA-CAD method is more convenient and cheaper than LC-MS/MS methods. The method was successfully used in a major project called "Special Action against Counterfeit and Shoddy white spirits" and to monitor risks posed by white spirits in China.

Keywords: sweeteners; photo-diode array detector (PDA); charged aerosol detection (CAD); white spirits

1. Introduction

Humans favor and instinctively desire sweet tastes, resulting in a preference for sweet foodstuffs [1,2]. Legal requirements and consumer pressure to more effectively monitor food safety (for dairy products, sweeteners, and alcoholic products) have recently become important in China. Chinese regulations for sweeteners are published in the national food safety standards for the use of food additives GB 2760 [3]. The maximum concentrations of certain sweeteners in specific types of food are contained in these standards. The standards are constantly revised to keep pace with technological developments in the sweetener field and to ensure that the maximum allowed concentrations of high-potency sweeteners in foods in specific categories are appropriate. Many food products contain sweeteners (singly or in combination), and it is essential that the concentrations of sweeteners in food products are below the maximum concentrations specified in the relevant legislation. The Chinese national food safety standards for wine GB 15037 and GB 2758 [4–6] prohibit acesulfame (ACS-K), aspartame (ASP), neotame (NEO), saccharin (SAC), and sodium cyclamate (CYC) (see Figure 1) being added to wine products. According to GB 2760, sucralose were authorized at concentration of 0.65 g/kg in fermented wine. Unfortunately, these sweeteners are often illegally added to various white spirits (which is a subdivision category of fermented wine). The presence of these sweeteners in such products may pose risks to human health, and the sweeteners may cause conditions, such as allergic reactions, bladder cancer, convulsions, hyperpnea, and metabolic acidosis [7,8]. However, some food producers may still add such sweeteners to their products and therefore cause health risks. The sweeteners mentioned above are the most common sweeteners that have been abused in recent years. It is therefore necessary to develop a rapid and accurate method to simultaneously determine the concentrations of these sweeteners in white spirits to ensure that the spirits comply with food safety standards and to assure the public, particularly in China, that spirits being sold met the needs of consumers.

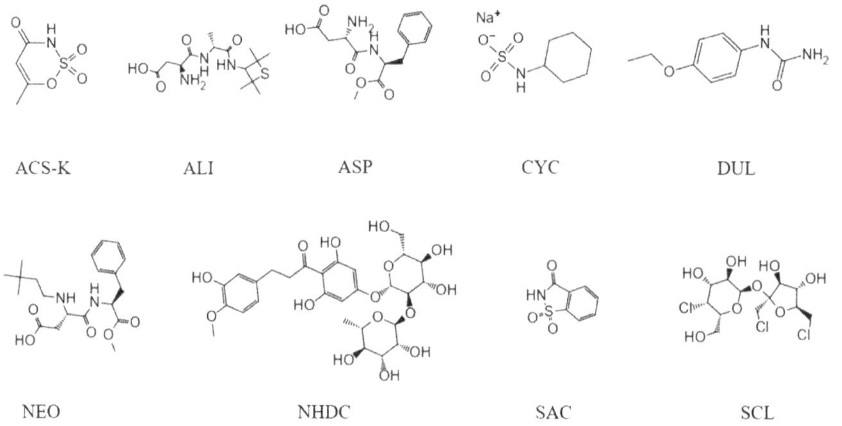

Figure 1. Chemical structures of the sweeteners that were studied.

Many "traditional" methods have been developed for determining high-intensity sweeteners in various foodstuffs. These methods are based on various analytical techniques, including high-performance liquid chromatography (HPLC) [9], ion chromatography [10], thin-layer chromatography [11], gas chromatography [12], capillary electrophoresis [13], flow injection analysis [14], electroanalytical techniques [15], nuclear magnetic resonance [16], and spectroscopic techniques [17]. However, most previously developed methods can only be used to analyze one sweetener or simple mixtures of two to four sweeteners. Nowadays, sweeteners are often used as synergistic mixtures to decrease costs and improve the taste of the product, and the maximum permissible amounts of different sweeteners in food vary markedly [18]. It is very important to have an analytical method available for simultaneously determining various sweeteners in various food matrices to allow food quality to be controlled and regulations to be enforced. Wasik et al. [19] developed an HPLC evaporative light-scattering detection method for determining six authorized sweeteners (ACS-K, ASP, CYC, neohesperidine dihydrochalcone (NHDC), SAC, and SCL). Yan et al. [20] used HPLC with evaporative light scattering detection analysis sucralose and related compounds with a better limit of detection (0.5 µg/mL). Koyama et al. [21] developed an LC mass spectrometry (MS) method for simultaneously determining nine sweeteners (ACS-K, ASP, CYC, dulcin (DUL), glycyrrhizic acid, rebaudioside A, SAC, SCL, and stevioside) in various foods. Koyama et al. and Huang [21,22] did not use internal standards (isotopic internal standard) in their methods, and the analytes were detected using the MS instruments in selected ion monitoring mode (SIM) would probably have better relative standard deviations (RSD) if using an isotopic internal standard. LC/tandem MS (MS/MS) is an increasingly popular technique for determining multiple sweeteners in wine and other alcoholic drinks [22–30]. Zygler et al. [24] developed an LC-MS/MS method to determine nine European Union-regulated sweeteners in alcoholic beverages. The internal standard was *N*-(2-methylcyclohexyl) sulfamate, and the limit of detection (LOD) was <0.5 μg/g. Chui-Shiang Chang et al. [27] developed an LC-MS/MS method to determine seven sweeteners in alcoholic beverages, and the LOD for each sweetener was 0.1 μg/g. However, LC-MS/MS methods are the most effective, in terms of quantifying the analytes, if isotope-labeled internal standards are used, meaning specially synthesized isotope-labeled sweeteners are required for LC-MS/MS analysis to be effective. LC-MS/MS methods often suffer from matrix effects and require complex sample processing procedures, However, simultaneous analysis of various food sweeteners, including ACS-K, ASP, CYC, NHDC, SAC, and SCL, in wine rapidly and convenience remain an area to be explored. So, LC-MS/MS analyses of sweeteners are more expensive (analysis cost and time) than the ultrahigh performance liquid chromatography (UHPLC) with photo-diode array (PDA) detection and the charged aerosol detection (CAD) method proposed here. Grembecka et al. [31] present a combination of HPLC-CAD-UV/DAD detectors to determine three sweeteners (ACS-K, ASP, and SAC) and two preservatives (citric acid and sodium benzoate) in soft drinks, which was a water soluble matrix. Fermented wine contains sweeteners like sulfonamides, dipeptides, and sucrose derivatives, as well as a complex matrix that is water soluble with chemical families, namely esters, alcohols, terpenic compounds, amino acid, and sulphur compounds, etc. [32]. It is desirable to develop cheap, simple, and fast methods for simultaneous analysis of various synthetic and semi-synthetic high-intensity sweeteners (e.g., ACS-K, alitame (ALI), ASP, CYC, DUL, NEO, NHDC, SAC, and SCL) in wine by HPLC combined with different detectors.

In the presented study, a new method for analyzing multiple sweeteners by UHPLC-PDA-CAD was developed and validated. The method was suitable for analyses to apply the maximum synthetic and semi-synthetic high-intensity sweetener concentrations in ethanol matrix, and make the current standard (food specified in Chinese legislation [3–6]) more sophisticated. The UHPLC separation method was optimized, and the effects of varying the method parameters on the recoveries, precision, linear range, limits of detection (LODs), and limits of quantification (LOQs) were assessed. The method was used to determine sweeteners in 30 real spirit samples. Finally, the method was successfully used in a major project called "Special Action against Counterfeit and Shoddy white spirits" and to monitor risks posed by white spirits in China.

2. Results and Discussion

2.1. Optimization of UHPLC Separation

The first step in developing the new method was selecting an analytical column. A rapid C18 column was selected. This allowed the nine selected sweeteners (ACS-K, ALI, ASP, CYC, DUL, NEO, NHDC, SAC, and SCL) to be eluted within 16 min. Columns of different types from various manufacturers were tested. The columns that were tested were a Shim-pack XR-C18 column (3.0 mm i.d., 75 mm long, 2.2 μm particle size), a Zorbax SB-C18 column (2.1 mm i.d., 50 mm long, 1.8 μm particle size), an Acquity UPLC BEH C18 column (2.1 mm i.d., 50 mm long, 1.7 μm particle size), and an Acquity UPLC BEH C18 column (2.1 mm i.d., 100 mm long, 1.7 μm particle size). As shown in Figure 2, good separation of the nine sweeteners was achieved using every column when the chromatographic conditions were optimized. For more information of the resolution and tailing factor of the nine sweeteners, see the Supplementary Materials in Table S3.

Several parameters (including the mobile phase and the gradient elution parameters) needed to be optimized. Firstly, two complementary detectors, a PDA and CAD, were found to be necessary. The main advantage of using the PDA and CAD in series was that CAD could detect sweeteners regardless of whether they contained chromophores or fluorophores, but the PDA detector could detect ACS-K, DUL, NHDC, and SAC with better sensitivity than that achieved by CAD. Each detector could be used to verify the results of the other detector. Secondly, a buffered mobile phase needed to be selected to give stable retention times. Most of the sweeteners could form charged molecules in polar mobile phases commonly used in reverse-phase LC systems. The degree of ionization of a molecule will affect interactions between the molecule and the stationary phase. The nine sweeteners were separated most effectively using methanol and 10 mmol/L ammonium acetate solution (at pH 3.8) as the mobile phases and using a gradient elution program. For more information on optimization of pH, mobile phase, and gradient conditions, please see the Supplementary Materials.

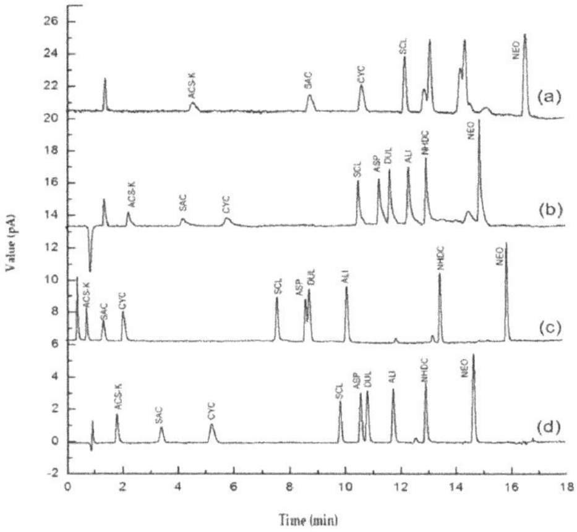

Figure 2. Ultrahigh performance liquid chromatography charged aerosol detector chromatograms for a mixture of nine sweeteners acquired using four different analytical columns, (**a**) a Shim-pack XR-C18 column (3.0 mm × 2.2 μm × 75 mm), (**b**) a Zorbax SB-C18 column (2.1 mm × 1.8 μm × 50 mm), (**c**) an Acquity UPLC BEH C18 column (2.1 mm × 1.7 μm × 50 mm), and (**d**) an Acquity UPLC BEH C18 column (2.1 mm × 1.7 μm × 100 mm).

2.2. Sample Preparation

2.2.1. Sample Preparation with Nitrogen Blowing

The samples were successfully prepared (i.e., the matrices were simplified) by evaporating them under a gentle stream of nitrogen. Due to the large amount of ethanol in liquor, it was easy to affect the retention performance of the sweetener in the process of liquid phase separation. Therefore, it was necessary to dealcoholize the wine samples to be tested in advance. In order to avoid the loss of other components in the sample caused by a temperature increase, we adopted the nitrogen blowing method at 35 °C. Figure 3 shows the workflow of the sample preparation with nitrogen blowing.

We also tried sample pretreatment without nitrogen and solid phase extraction (SPE). The processes of sample pretreatment without nitrogen was basically the same as treatment with nitrogen except the second part. The regent tube was put into the water bath and did not inject nitrogen. For the SPE steps, please see the Supplementary Materials.

Evaporating a sample under a gentle stream of nitrogen was a more effective treatment than solid phase extraction. It took ~25 min to prepare a sample by evaporating it under a gentle stream of nitrogen, and the cost was ~3.0 yuan per sample. Figure 4-(a) shows that ACS-K, SAC, and CYC could hardly be detected, and the chromatographic peak of SCL was seriously deformed. The solid phase extraction procedure took 55 min using a fully automatic solid phase extraction unit, and cost ~8.0 yuan per sample (for the extraction column, solvent, and other materials). Figure 4-(b) (for a sample that was evaporated under a gentle stream of nitrogen) shows a low baseline and good symmetry. Preparing a sample by evaporating it under a gentle stream of nitrogen was simple, fast, and cheap.

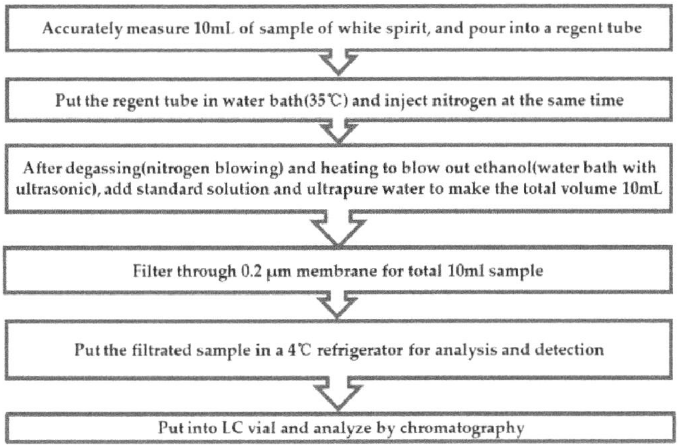

Figure 3. Sample preparation with nitrogen blowing.

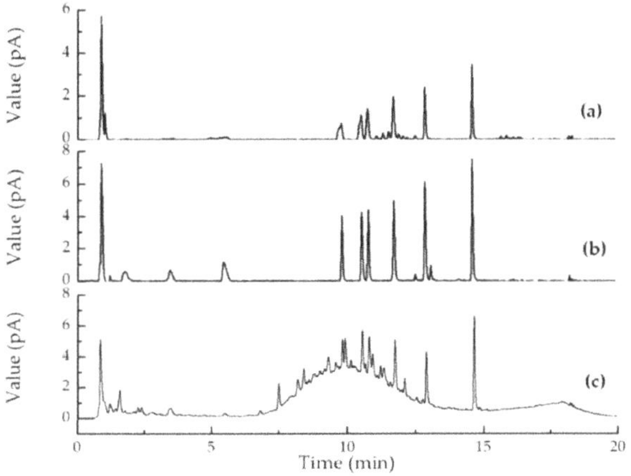

Figure 4. Chromatograms of the pretreatment for spirit samples (white spirit samples added 10 μg/g of sweetener solution and treated respectively (**a**) without the pretreatment of nitrogen, (**b**) evaporation under a gentle stream of nitrogen, and (**c**) solid phase extraction treatment).

2.2.2. Selection of Filter Membrane

We also analyzed the microporous filter membrane, which significantly affected the preparation results. The mixed standard solution used in the experiment needed to be processed by a vortex oscillator before detection to ensure that the sweetener was fully dissolved. However, there were still some particles or impurities in it. If the sample was injected directly, it was easy to block the chromatographic column and shorten the life of the column; at the same time, it also had a certain impact on the detector. Therefore, it was necessary to filter the samples to be tested with a microporous filter membrane to protect the components, such as the column and inject valve, from pollution. The membrane used in the experiment was investigated and analyzed, as shown in Figure 5.

We compared four microporous membranes. The blank water was treated with nylon, PVDF, PTFE, and PES, respectively, then directly injected in UHPLC-CAD. The results showed some impurity peaks after about 10 min of nylon, PVDF, and PTFE. The probable reason was that a small amount of chemical substances on the membrane fell off, which could also be detected. The modified PES microporous membrane had chemical and thermal stability, and acid and alkali resistance (pH 1–14), thus ensuring low dissolution and good reproducibility. Through comparison, the PES microporous membrane was used to treat the mixed standard solution and the real sample.

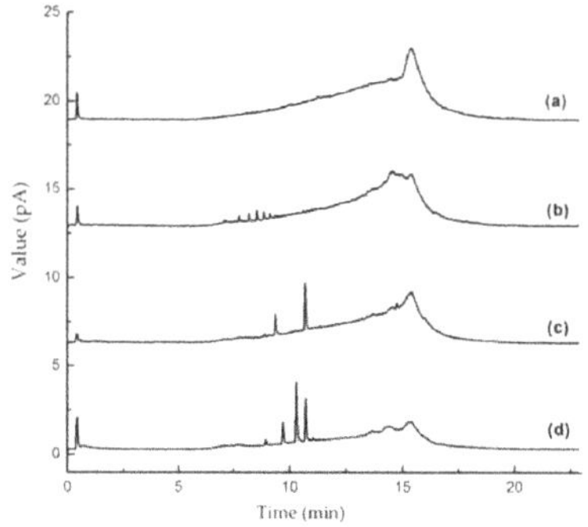

Figure 5. Chromatograms of blank water detected by four kinds of membranes. ((**a**) Polyethersulfone microporous membrane (PES) 0.20 μm, (**b**) Nylon microporous membrane, 0.20 μm, (**c**) polyvinylidene fluoride (PVDF) membrane 0.20 μm, (**d**) Polytetrafluoroethylene membrane (PTFE) 0.20 μm).

2.3. Method Validation and Application of the Method to Real Samples

2.3.1. UHPLC-PDA-CAD Chromatogram

The nine sweeteners were quantified by analyzing calibration standard solutions using the same UHPLC conditions that were used for the white spirits. The nine sweeteners were effectively separated within 16 min using the optimum conditions. Typical UHPLC-PDA-CAD chromatograms for the target compounds are shown in Figure 6. All the analytes were determined simultaneously by UHPLC-PDA-CAD using the rapid column. The chromatograms indicated that the UHPLC resolution and peak shapes were acceptable. Four compounds (including ACS-K, DUL, NHDC, and SAC) were identified in the PDA chromatogram acquired at a wavelength of 226 nm. All the sweeteners were identified in the CAD chromatogram and were able to be quantified.

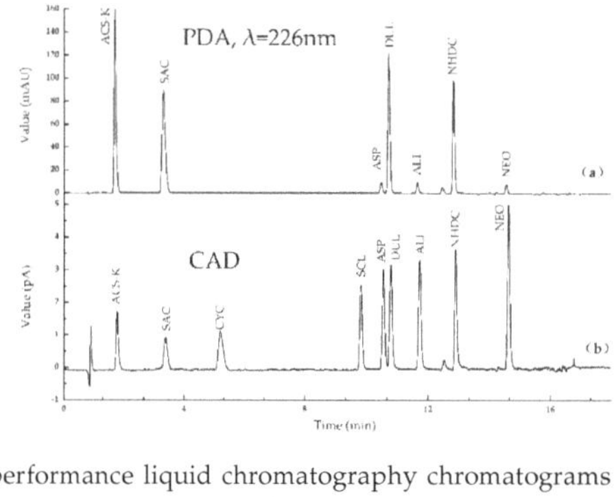

Figure 6. Ultrahigh performance liquid chromatography chromatograms for a standard solution containing each sweetener at a concentration of 10 μg/g acquired using the photo-diode array detector ((**a**), using a wavelength of 226 nm) and using the charged aerosol detector (**b**).

All 30 products from 12 different brands were collected from different areas in Beijing supermarkets (China). All the samples were stored under refrigeration conditions (4 °C) until analysis. Because all the sweeteners in the present study had good solubility in water, the food samples could be directly used for sweetener analysis. The samples were treated using the nitrogen blow method, adjusted the pH to 3.8, and filtered through a 0.20μm syringe filter prior to being injected into UHPLC-PDA-CAD for analysis.

2.3.2. Linear Ranges, Regression Equations, the Limit of Detection (LOD), the Limit of Quantization (LOQ), Repeatability and Reproducibility

The linear ranges of the calibration curves for the sweeteners were determined. For the CAD data, the calibrations were linear over the concentration range 0.5–50.0μg/g, and the coefficients of determination (γ^2) were 0.9937 to 0.9963. The repeatability was determined by analyzing standard solutions containing the sweeteners, each at a concentration of 5.0μg/g, and the results are shown in Table 1. The repeatability for the sweeteners was 1.2% to 3.1% (CAD) and the reproducibility was 2.3% to 3.6% (CAD). These results indicated that the method was precise and fit for purpose. The LOD and LOQ were defined as the concentrations giving signal-to-noise ratios of 3 and 10, respectively. The LOD and LOQ for the sweeteners are shown in Table 1. See the Supplementary Materials for the linearity with PDA and CAD (Supplementary Materials S3, S4).

The LOD for the CAD data were 0.36 μg/g for ACS-K, 0.19 μg/g for ALI, 0.20 μg/g for ASP, 0.32 μg/g for CYC, 0.18 μg/g for DUL, 0.16 μg/g for NEO, 0.16 μg/g for NHDC, 0.77 μg/g for SAC, and 0.18 μg/g for SCL. The LOD for the PDA data were 0.16 μg/g for ACS-K, 0.18 μg/g for DUL, 0.21 μg/g for NHDC, and 0.22 μg/g for SAC. These LOD were much lower than those found in previous studies [9,19–22,31] for methods involving HPLC evaporative light-scattering detection and HPLC-MS (13.0 μg/g for ACS-K, 2.0 μg/g for ALI, 10.0 μg/g for ASP, 1.0 μg/g for CYC, 6.0 μg/g for DUL, 5.0 μg/g for NEO, 2.0 μg/g for NHDC, 2.0 μg/g for SAC, and 1.0 μg/g for SCL). This clearly indicates that the method described here was very effective for analyzing sweeteners in white spirits. This was particularly the case for ALI, NHDC, and SCL, which had LOD much lower than required. Table 2 compares the LOD and LOQ in the related literature.

2.3.3. Recoveries and Accuracy

The accuracy of the method was assessed by analyzing three white spirit samples. The samples contained 38°, 46°, and 52° white spirits (calculated with ethanol). Each sample was spiked with the nine sweeteners at fortification levels (for each sweetener) of 5.0, 20.0, and 40.0 μg/g. The recoveries are presented in Table 3. Good recoveries were obtained for the 38° spirit samples (95.9–104.5% recoveries), the 46° spirit samples (95.1–103.8% recoveries), and the 52° spirit samples (95.5–104.9% recoveries). As shown in Table 3, the precisions were 2.5% to 3.8% for the 38° spirit samples, 2.2% to 3.4% for the 46° spirit samples, and 1.6% to 3.0% for the 52° spirit samples.

Table 1. Chromatographic data, linear ranges, regression equations [a], correlation coefficients, limits of detection [b], limits of quantitation [c], repeatability [d], and reproducibility [e] for the nine sweeteners in white spirits using the ultrahigh performance liquid chromatograph photo-diode array detector and charged aerosol detector method.

Analytes	$t_R \pm S$ (min)	CAD								PDA (λ=226nm)					
		Resolution	Linear Ranges [f]	Linear Equation	γ^2	LOD	LOQ	Repeatability	Reproducibility	Linear Equation	γ^2	LOD	LOQ	Repeatability	Reproducibility
ACS-K	1.79 ± 0.09	8.67	1.0–50.0	y = 0.0124 x + 0.0212	0.9946	0.36	1.06	2.4%	3.1%	y = 1.275 *x* − 0.124	0.9998	0.16	0.50	1.1%	1.2%
SAC	3.35 ± 0.02	6.89	2.0–50.0	y = 0.0109 x + 0.0067	0.9937	0.77	2.07	2.1%	3.0%	y = 1.183 *x* − 0.188	0.9999	0.22	0.60	0.9%	1.3%
CYC	5.19 ± 0.03	19.9	1.0–50.0	y = 0.0207 x + 0.0318	0.9963	0.32	0.95	3.1%	3.6%	/ [g]	/	/	/	/	/
SCL	9.82 ± 0.03	4.68	0.7–50.0	y = 0.0202 x + 0.0446	0.9949	0.18	0.52	2.0%	2.3%	/	/	/	/	/	/
ASP	10.55 ± 0.03	1.58	0.7-50.0	y = 0.0237 x + 0.0523	0.9942	0.20	0.59	1.2%	2.9%	/	/	/	/	/	/
DUL	10.80 ± 0.04	5.84	0.5-50.0	y = 0.0265 x + 0.0524	0.9956	0.18	0.54	2.0%	3.4%	y = 1.056 *x* − 0.050	0.9999	0.18	0.50	0.8%	1.6%
ALI	11.72 ± 0.03	7.64	0.5–50.0	y = 0.0285 x + 0.0531	0.9957	0.19	0.59	1.4%	3.1%	/	/	/	/	/	/
NHDC	12.91 ± 0.07	11.9	0.5–50.0	y = 0.0313 x + 0.0427	0.9963	0.16	0.53	1.6%	2.9%	y = 0.771 *x* − 0.288	0.9993	0.21	1.01	1.0%	1.7%
NEO	14.64 ± 0.06	3.51	0.5–50.0	y = 0.0425 x + 0.0714	0.9953	0.16	0.49	1.5%	3.0%	/	/	/	/	/	/

Notes: [a] calibration curves with different compounds at concentrations of 0.5, 1.0, 2.0, 5.0, 10.0, 20.0, and 50.0 μg/g. [b] the limit of detection (LOD) was evaluated based on a signal-to-noise ratio of 3 (*S/N*), unit: μg/g. [c] the limit of quantization (LOQ) was evaluated based on a signal-to-noise ratio of 10 (*S/N*), unit: μg/g. [d] the repeatability (n = 7). Compound concentration at 5.0 μg/g. [e] the reproducibility (n = 5, two analysts, 2 times/day). Compound concentration at 5.0 μg/g. [f] the liner range unit: μg/g. [g] not available, as CYC, SCL, ASP, ALI, NEO did not respond at 226 nm (PDA).

Table 2. Comparison of the limit of detection(LOD) and limit of quantization(LOQ) in the related literatures.

	Matrices	Analytes	LOD	LOQ	Ref
UPLC-UV	juices	ACS-K ASP SAC	0.75 μg/mL 0.75 μg/mL 0.30 μg/mL	NA	[9]
HPLC-ELSD	canned fruits, yoghurt, energy drink	ACS-K ALI ASP CYC DUL NHDC NEO SAC SCL	13.0 μg/g 2.0 μg/g 10.0 μg/g 1.0 μg/g 6.0 μg/g 2.0 μg/g 5.0 μg/g 2.0 μg/g 1.0 μg/g	NA	[19]
HPLC-ELSD	commercial samples	SCL	0.5 μg/mL	2.0 μg/mL	[20]
LCMS	food	ACS-K SCL SAC CYC ASP DUL	NA	1–5 μg/g	[21]
LCMS (ion-pair)	food	CYC	1 ng/mL	5 ng/mL	[22]
HPLC-CAD-UV/DAD	soft drinks	ASP ACS-K SAC	0.08–0.20 μg/mL	0.19–0.61 μg/mL	[32]
UHPLC-PDA-CAD	white spirits	ACS-K ALI ASP CYC DUL NHDC NEO SAC SCL	0.36 μg/g 0.19 μg/g 0.20 μg/g 0.32 μg/g 0.18 μg/g 0.16 μg/g 0.16 μg/g 0.77 μg/g 0.18 μg/g	1.06 μg/g 0.59 μg/g 0.59 μg/g 0.95 μg/g 0.54 μg/g 0.53 μg/g 0.49 μg/g 2.07 μg/g 0.52 μg/g	Present method

Table 3. Recovery and accuracy results for determining nine sweeteners in three different spirit samples.

Analytes	Added (μg/g)	White Spirits 38°			White Spirits 46°			White Spirits 52°		
		Found	Recovery	RSD	Found	Recovery	RSD	Found	Recovery	RSD
ACS-K	5.0	4.88	97.6%	3.0%	4.74	94.8%	2.8%	4.80	96.0%	2.5%
	10.0	9.59	95.9%	3.2%	9.67	96.7%	2.7%	9.71	97.1%	2.2%
	40.0	38.49	96.2%	2.9%	38.18	95.4%	2.5%	39.04	97.6%	2.1%
CYC	5.0	5.23	104.5%	3.6%	5.12	102.5%	3.4%	5.25	104.9%	2.9%
	10.0	10.20	102.0%	3.8%	10.28	102.8%	3.2%	10.21	102.1%	3.0%
	40.0	40.32	100.8%	3.1%	41.15	102.9%	2.9%	41.56	103.9%	2.6%
SAC	5.0	4.80	96.0%	3.1%	4.76	95.1%	2.8%	4.78	95.5%	2.3%
	10.0	9.63	96.3%	3.0%	9.78	97.8%	2.6%	9.69	96.9%	2.2%
	40.0	39.64	99.1%	3.3%	40.33	100.8%	3.0%	40.25	100.6%	2.3%
SCL	5.0	4.99	99.8%	2.9%	5.04	100.8%	2.2%	4.98	99.5%	2.3%
	10.0	9.96	99.6%	2.7%	10.02	100.2%	2.8%	9.95	99.5%	2.6%
	40.0	40.33	100.8%	3.1%	39.88	99.7%	2.7%	39.67	99.2%	2.4%
ASP	5.0	4.98	99.7%	2.8%	5.13	102.6%	2.3%	4.90	98.1%	1.6%
	10.0	9.97	99.7%	2.5%	9.81	98.1%	2.4%	9.89	98.9%	1.9%
	40.0	39.00	97.5%	2.7%	38.62	96.5%	2.4%	39.24	98.1%	2.0%
DUL	5.0	4.86	97.1%	3.3%	4.72	94.3%	2.3%	4.82	96.3%	2.7%
	10.0	10.01	100.1%	3.0%	9.82	98.2%	2.9%	9.62	96.2%	2.6%
	40.0	38.56	98.4%	3.2%	38.80	97.0%	2.8%	39.93	99.8%	2.4%
ALI	5.0	4.79	98.8%	3.3%	4.76	98.2%	3.0%	4.76	97.2%	2.4%
	10.0	9.90	96.0%	3.1%	9.97	97.7%	2.6%	9.77	97.7%	1.9%
	40.0	38.76	97.0%	3.4%	39.07	97.7%	2.9%	38.88	97.3%	2.3%
NHDC	5.0	4.89	97.8%	3.1%	4.81	96.2%	2.2%	4.90	98.1%	1.9%
	10.0	9.88	98.8%	3.0%	9.97	99.7%	2.5%	9.93	99.3%	2.2%
	40.0	39.39	98.5%	3.3%	39.01	97.5%	2.6%	39.08	97.7%	2.3%
NEO	5.0	4.91	98.2%	2.9%	4.79	95.9%	2.6%	4.96	99.1%	1.8%
	10.0	9.69	96.9%	3.3%	9.81	98.1%	2.7%	9.82	98.2%	2.1%
	40.0	38.51	96.3%	3.0%	38.74	96.9%	2.8%	38.71	96.8%	2.3%

2.4. Real Sample Analysis

The concentrations of the nine sweeteners in 30 real white spirit samples (12 brands), containing either 38°, 46°, or 52° alcohol were determined (Supplementary Materials Table S1). Three samples were found to contain sweeteners. Sample 1 (a) contained SCL at a concentration of 8.45μg/g (determined using the CAD data). Sample 7 (b) contained NEO at a concentration of 1.07μg/g (determined using the CAD data). Sample 12 (c) contained SAC at a concentration of 3.22μg/g (determined using the PDA data). The chromatograms for these samples are shown in Figure 7. The LOD allowed the low concentrations of the sweeteners in the white spirit samples to be determined. The artificial sweetener content contained in sample No.1 was outside of the legal limit according to GB 2760.

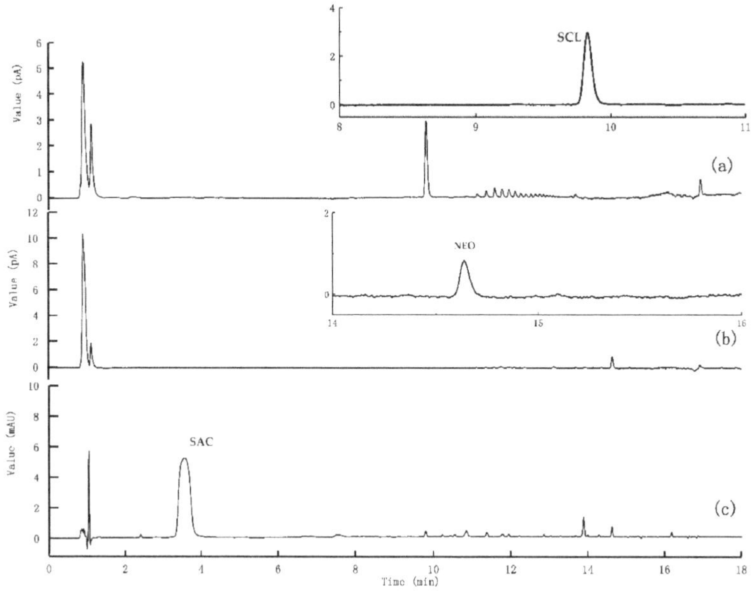

Figure 7. Ultrahigh performance liquid chromatography photo-diode array detector and charged aerosol detector chromatograms for samples that were found to contain sweeteners.

3. Materials and Methods

3.1. Instrumentation and Reagents

Chromatograms were acquired using a UHPLC system consisting of two GP40 LC pumps, an AS50 autosampler, an LC 20 column compartment, an Ultimate 3000 PDA detector, and a Corona Ultra CAD, with the detectors linked using a series connection (Thermo Fisher Scientific, Waltham, MA, USA). Nine food sweetener reference materials were provided by the National Institute for Metrology of China (Beijing, China). The sweetener reference materials (RM) were ACS-K (GBW(E)100065, 99.6% ± 0.6% pure), ALI (98.3% ± 0.7% pure), ASP (99.0% ± 0.7% pure), CYC (GBW(E)100066, 99.3% ± 0.7% pure), DUL (98.0% ± 1.0% pure), NEO (98.7% ± 0.6% pure), NHDC (99.0% ± 0.6% pure), SAC (GBW10006, 99.5% ± 0.5% pure), and SCL (GBW(E)100132, 99.6% ± 0.5% pure). HPLC-grade ammonium acetate and formic acid were purchased from Sigma-Aldrich (St. Louis, MO, USA). Methanol and acetonitrile (HPLC grade) were purchased from Merck (Darmstadt, Germany). Further, 0.20μm, polyethersulfone(PES), nylon, polyvinylidene fluoride (PVDF) and polytetrafluoroethylene(PTFE) four membranes (Shanghai ANPEL Scientific Instrument Co., Ltd.) adsorbents were provided by Agela Technologies company (Tianjin, China). Automatic solid phase extractor, Fotector-06C, Reeko (Xiamen, China). Solid phase extraction column, Strata-X 33μm (3 mL/200 mg), Chromabond C18 (6 mL/1000 mg), Agela Technologies company (Tianjin, China).

3.2. Preparation of Standard Solutions and Samples

The standard solutions were prepared on a weight–weight basis. ACS-K, CYC, SAC, and SCL solutions were each prepared by dissolving an aliquot of the relevant sweetener in deionized water to give a solution containing the sweetener at a concentration of 500 μg/g. ALI, ASP, and NEO solutions were each prepared by dissolving an aliquot of the relevant sweetener in 10 mmol/L ammonium acetate solution to give a solution containing the sweetener at a concentration of 500 μg/g. DUL and NHDC solutions were each prepared by dissolving an aliquot of the relevant sweetener in deionized water to give a solution containing the sweetener at a concentration of 100 μg/g. Working mixed standard solutions at concentrations lower than 1.0 μg/g were prepared by diluting 1.0 μg/g mixed standard solutions as required. Standard solutions of the sweeteners at concentrations between 0.5 and 50.0 μg/g were prepared by performing serial dilutions starting with high concentration standard solutions. In total, 12 brand real white spirits for 30 samples, were obtained from local markets.

3.3. UHPLC-PDA-CAD Conditions

Separation of the sweeteners was achieved using an Acquity UHPLC BEH C18 column (2.1 mm i.d., 100 mm long, 1.7-μm particle size, Waters Corporation, made in Ireland) using a gradient profile. The mobile phases were (A) methanol and (B) 10 mmol/L ammonium acetate solution, and the gradient elution program started at 10% A, which was held between 0 and 3 min, increased linearly to 90% A between 3 and 13 min, remained at 90% A until 15 min, and then returned to 10% A. The mobile phase flow rate was 0.3 mL/min. The column oven temperature was 35 °C. The injection volume was 5 μL. The PDA detector monitoring wavelength range was 190–700 nm, The CAD detector nebulizer temperature was 35 °C, the gas pressure was 0.24 MPa, and the data collection rate was 20 Hz.

3.4. Method Validation

Quantitative analysis was performed using an external standards calibration method. The calibration solutions were prepared by diluting intermediate mixed aqueous standard solutions to give sweetener concentrations between 0.5 and 50.0 μg/g. The LOD and LOQ were defined as the concentrations giving signal-to-noise ratios of 3 and 10, respectively. Repeatability (intra-day precision) was assessed by analyzing a standard solution containing the sweeteners each at a concentration of 5.0 μg/g seven times in one day. Reproducibility (inter-day precision) was assessed by two different analysts analyzing the same standard twice each day for five days. Accuracy was assessed by performing recovery experiments using three wine samples with different origins spiked with the nine sweeteners at concentrations of 5.0, 10.0, and 40.0 μg/g. Each spiked sample was analyzed seven times. Matrix effects were assessed by adding standard sweetener solutions to a blank white spirit sample to give final sweetener concentrations of 5.0 and 20.0 μg/g and then analyzing the samples.

4. Conclusions

An efficient and accurate method was developed for simultaneously determining nine sweeteners in Chinese spirits by UHPLC-PDA-CAD. The method included a simple sample preparation procedure (evaporation under a gentle stream of nitrogen and filtration), and is sensitive, cheap, simple, and quick. The method allows synthetic and semi-synthetic high-intensity sweeteners to be detected at low concentrations (micrograms per gram). The applicability of the method was verified by determining sweeteners in 30 real spirit samples. Finally, the method was successfully used in a major project called "Special Action against Counterfeit and Shoddy white spirits" and to monitor risks posed by white spirits in China.

Supplementary Materials: The following are available online at http://www.mdpi.com/1420-3049/25/1/40/s1, Figure S1: Area differentiation of nine sweeteners adjusting pH with different concentration of ammonium acetate, Figure S2: The effect of different mobile flows on separation, Figure S3: Linearity of nine sweeteners(PDA), Figure S4: Linearity of nine sweeteners (CAD), Figure S5: Optimization of pretreatment conditions, Figure S6: Workflows of three preparation methods, Table S1: Three of the thirty samples detected illegally added sweetener, Table S2: Operation order list of solid phase extraction, Table S3: The Resolution and Tailing factor of nine sweeteners by UHPLC-CAD, Table S4: The separation gradient condition for nine sweeteners.

Author Contributions: K.M., X.L. and Y.Z. did the experiments, analyzed the data, and wrote the paper. K.M. and F.L. conceived, designed the experiment, and elaborated the paper, and all authors approved the final paper. All authors have read and agreed to the published version of the manuscript.

Funding: This research was funded by the National Key R&D Program of China, Research and Application of common Technology of National Quality Infrastructure Project (No 2018YFF0212800 and 2018YFF0212801) and the General Administration of Quality Supervision, Inspection and Quarantine of the People's Republic of China (No. 2012104001).

Acknowledgments: We thank Gareth Thomas, PhD, from Edanz Group (www.edanzediting.com/ac) for editing a draft of this manuscript.

Conflicts of Interest: The authors declare no conflict of interest. The funders had no role in the design of the study; in the collection, analyses, or interpretation of data; in the writing of the manuscript, or in the decision to publish the results.

References

1. DuBois, G.E.; Prakash, I. Non-Caloric Sweeteners, Sweetness Modulators, and Sweetener Enhancers. *Annu. Rev. Food Sci. Technol.* **2012**, *3*, 353–380. [CrossRef]
2. Carocho, M.; Morales, P.; Ferreira, I.C.F.R. Sweeteners as Food Additives in the Xxi Century: A Review of What Is Known, and What Is to Come. *Food Chem. Toxicol.* **2017**, *107*, 302–317. [CrossRef]
3. Ministry of Health of the People's Republic of China; National Institute of Standards of the People's Republic of China. *National Standard of the People's Republic of China GB7718-2004 of General Standard for the Labeling of Prepackaged Foods*; Standard Press of China: Beijing, China, 2004.
4. Ministry of Health of the People's Republic of China; National Institute of Standards of the People's Republic of China. *National Standard of the People's Republic of China GB2760-2014 of Hygienic Standards for Uses of Food Additives*; Standard Press of China: Beijing, China, 2014.
5. Ministry of Health of the People's Republic of China; National Institute of Standards of the People's Republic of China. *National Standard of the People's Republic of China GB 15037-2006 of Wine*; Standard Press of China: Beijing, China, 2006.
6. Ministry of Health of the People's Republic of China; National Institute of Standards of the People's Republic of China. *National Standard of the People's Republic of China GB2758-2012 of Hygienic Standard for Fermented Alcoholic Beverages*; Standard Press of China: Beijing, China, 2012.
7. Praveena, S.M.; Cheema, M.S.; Guo, H.-R. Non-Nutritive Artificial Sweeteners as an Emerging Contaminant in Environment: A Global Review and Risks Perspectives. *Ecotox. Environ. Saf.* **2019**, *170*, 699–707. [CrossRef]
8. Luo, J.; Wu, L.; Zhang, Q.; Wu, Y.; Fang, F.; Feng, Q.; Li, C.; Xue, Z.; Cao, J. Review on the Determination and Distribution Patterns of a Widespread Contaminant Artificial Sweetener in the Environment. *Environ. Sci. Pollut. Res.* **2019**, *26*, 19078–19096. [CrossRef] [PubMed]
9. Tighrine, A.; Amir, Y.; Alfaro, P.; Mamou, M.; Nerin, C. Simultaneous Extraction and Analysis of reservatives and Artificial Sweeteners in Juices by Salting out Liquid-Liquid Extraction Method Prior to Ultra-High Performance Liquid Chromatography. *Food Chem.* **2019**, *277*, 586–594. [CrossRef] [PubMed]
10. Zhu, Y.; Guo, Y.Y.; Ye, M.L.; James, F.S. Separation and Simultaneous Determination of Four Artificial Sweeteners in Food and Beverages by Ion Chromatography. *J. Chromatogr. A* **2005**, *1085*, 143–146. [CrossRef] [PubMed]
11. Nambiar, A.P.; Sanyal, M.; Shrivastav, P.S. Simultaneous Densitometric Determination of Eight Food Colors and Four Sweeteners in Candies, Jellies, Beverages and Pharmaceuticals by Normal-Phase High Performance Thin-Layer Chromatography Using a Single Elution Protocol. *J. Chromatogr. A* **2018**, *1572*, 152–161. [CrossRef]
12. Qiu, W.; Wang, Z.; Nie, W.; Guo, Y.; Huang, L. Gc-Ms Determination of Sucralose in Splenda. *Chromatographia* **2007**, *66*, 935–939. [CrossRef]

13. Bergamo, A.B.; Fracassi da Silva, J.A.; de Jesus, D.P. Simultaneous Determination of Aspartame, Cyclamate, Saccharin and Acesulfame-K in Soft Drinks and Tabletop Sweetener Formulations by Capillary Electrophoresis with Capacitively Coupled Contactless Conductivity Detection. *Food Chem.* **2011**, *124*, 1714–1717. [CrossRef]
14. Garcia-Jimenez, J.F.; Valencia, M.C.; Capitan-Vallvey, L.F. Simultaneous Determination of Antioxidants, Preservatives and Sweetener Additives in Food and Cosmetics by Flow Injection Analysis Coupled to a Monolithic Column. *Anal. Chim. Acta* **2007**, *594*, 226–233. [CrossRef]
15. Santini, A.O.; Lemos, S.C.; Pezza, H.R.; Carloni-Filho, J.; Pezza, L. Development of a Potentiometric Sensor for the Determination of Saccharin in Instant Tea Powders, Diet Soft Drinks and Strawberry Dietetic Jam. *Microchem. J.* **2008**, *90*, 124–128. [CrossRef]
16. Paniagua-Vega, D.; Cavazos-Rocha, N.; Huerta-Heredia, A.A.; Parra-Naranjo, A.; Rivas-Galindo, V.M.; Waksman, N.; Saucedo, A.L. A Validated NMR Method for the Quantitative Determination of Rebaudioside a in Commercial Sweeteners. *J. Food Compos. Anal.* **2019**, *79*, 134–142. [CrossRef]
17. Wang, Y.-T.; Li, B.; Xu, X.-J.; Ren, H.-B.; Yin, J.-Y.; Zhu, H.; Zhang, Y.-H. Ftir Spectroscopy Coupled with Machine Learning Approaches as a Rapid Tool for Identification and Quantification of Artificial Sweeteners. *Food Chem.* **2020**, *303*, 125404–125415. [CrossRef] [PubMed]
18. Martins, F.C.O.L.; Sentanin, M.A.; De Souza, D. Analytical Methods in Food Additives Determination: Compounds with Functional Applications. *Food Chem.* **2019**, *272*, 732–750. [CrossRef] [PubMed]
19. Wasik, A.; McCourt, J.; Buchgraber, M. Simultaneous Determination of Nine Intense Sweeteners in Foodstuffs by High Performance Liquid Chromatography and Evaporative Light Scattering Detection—Development and Single-Laboratory Validation. *J. Chromatogr. A* **2007**, *1157*, 187–196. [CrossRef] [PubMed]
20. Yan, W.; Wang, N.; Zhang, P.; Zhang, J.; Wu, S.; Zhu, Y. Simultaneous Determination of Sucralose and Related Compounds by High-Performance Liquid Chromatography with Evaporative Light Scattering Detection. *Food Chem.* **2016**, *204*, 358–364. [CrossRef]
21. Koyama, M.; Yoshida, K.; Uchibori, N.; Wada, I.; Akiyama, K.; Sasaki, T. Analysis of Nine Kinds of Sweeteners in Foods by LC/MS. *J. Food Hyg. Soc. Jap.* **2005**, *46*, 72–78. [CrossRef]
22. Huang, Z.Q.; Ma, J.Y.; Chen, B.; Zhang, Y.; Yao, S.Z. Determination of Cyclamate in Foods by High Performance Liquid Chromatography-Electrospray Ionization Mass Spectrometry. *Anal. Chim. Acta* **2006**, *555*, 233–237. [CrossRef]
23. Chen, X.H.; Zhao, Y.G.; Shen, H.Y.; Jin, M.C. Application of Dispersive Solid-Phase Extraction and Ultra-Fast Liquid Chromatography-Tandem Quadrupole Mass Spectrometry in Food Additive Residue Analysis of Red Wine. *J. Chromatogr. A* **2012**, *1263*, 34–42. [CrossRef]
24. Zygler, A.; Wasik, A.; Kot-Wasik, A.; Namiesnik, J. Determination of Nine High-Intensity Sweeteners in Various Foods by High-Performance Liquid Chromatography with Mass Spectrometric Detection. *Anal. Bioanal. Chem.* **2011**, *400*, 2159–2172. [CrossRef]
25. Zhao, Y.G.; Cai, M.Q.; Chen, X.H.; Pan, S.D.; Yao, S.S.; Jin, M.C. Analysis of Nine Food Additives in Wine by Dispersive Solid-Phase Extraction and Reversed-Phase High Performance Liquid Chromatography. *Food Res. Int.* **2013**, *52*, 350–358. [CrossRef]
26. Ma, K.; Li, X.J.; Wang, H.F.; Zhao, M. Rapid and Sensitive Method for the Determination of Eight Food Additives in Red Wine by Ultra-Performance Liquid Chromatography Tandem Mass Spectrometry. *Food Anal. Method* **2015**, *8*, 203–212. [CrossRef]
27. Chang, C.S.; Yeh, T.S. Detection of 10 Sweeteners in Various Foods by Liquid Chromatography/Tandem Mass Spectrometry. *J. Food Drug Anal.* **2014**, *22*, 318–328. [CrossRef] [PubMed]
28. Kubica, P.; Namiesnik, J.; Wasik, A. Comparison of Hydrophilic Interaction and Reversed Phase Liquid Chromatography Coupled with Tandem Mass Spectrometry for the Determination of Eight Artificial Sweeteners and Common Steviol Glycosides in Popular Beverages. *J. Pharm. Biomed. Anal.* **2016**, *127*, 184–192. [CrossRef] [PubMed]
29. Iwakoshi, K.; Tahara, S.; Uematsu, Y.; Yamajima, Y.; Miyakawa, H.; Monma, K.; Kobayashi, C.; Takano, I. Development of a Highly Sensitive Liquid Chromatography with Tandem Mass Spectrometry Method for the Qualitative and Quantitative Analysis of High-Intensity Sweeteners in Processed Foods. *J. Chromatogr. A* **2019**, *1592*, 64–70. [CrossRef] [PubMed]
30. Zygler, A.; Wasik, A.; Namiesnik, J. Analytical Methodologies for Determination of Artificial Sweeteners in Foodstuffs. *Trac-Trend Anal. Chem.* **2009**, *28*, 1082–1102. [CrossRef]

31. Grembecka, M.; Baran, P.; Blazewicz, A.; Fijalek, Z.; Szefer, P. Simultaneous Determination of Aspartame, Acesulfame-K, Saccharin, Citric Acid and Sodium Benzoate in Various Food Products Using HPLC -CAD-UV/Dad. *Eur. Food Res. Technol.* **2014**, *238*, 357–365. [CrossRef]
32. Perestrelo, R.; Silva, C.; Camara, J.S. Madeira Wine Volatile Profile. A Platform to Establish Madeira Wine Aroma Descriptors. *Molecules* **2019**, *24*, 3028. [CrossRef]

MDPI
St. Alban-Anlage 66
4052 Basel
Switzerland
Tel. +41 61 683 77 34
Fax +41 61 302 89 18
www.mdpi.com

Molecules Editorial Office
E-mail: molecules@mdpi.com
www.mdpi.com/journal/molecules

Lightning Source UK Ltd.
Milton Keynes UK
UKHW050032311222
414587UK00004B/71

9 783039 287345